Key Concepts in Life Sciences

KEY CONCEPTS IN LIFE SCIENCES

Edited by:

Akansha Singh

Delve
PUBLISHING

www.delvepublishing.com

Key Concepts in Life Sciences

Akansha Singh

Delve Publishing
224 Shoreacres Road
Burlington, ON L7L 2H2
Canada
www.delvepublishing.com
Email: orders@arclereducation.com

ISBN: 978-1-77469-380-3 (Hardcover)

ABOUT THE EDITOR

Dr. Akansha Singh is presently working as Project Scientist in Department of Genetics and Plant Breeding, Institute of Agricultural Sciences, Banaras Hindu University, India. She has also worked as Postdoctoral fellow, Mumbai University and as Associate Professor, Department of Genetics and Plant Breeding, College of Agriculture, Parul University, India She has obtained her Ph.D. (Ag) in Genetics and Plant breeding from Banaras Hindu University in the year 2012. She has been Awarded with ICAR, senior research fellowship in 2010 to pursue PhD. She has authored numerous national and international publications in journals of repute.

TABLE OF CONTENTS

PREFACE

'Life Science' is defined as the field of study of life and living things (also known as organisms). Usually, life sciences are referred to as biology. Life is amazingly complex and thus there are millions of species alive today. Many millions lived in the past and then went extinct. Clearly, life science is a huge science.

Life science is an enormous scientific field that intends to answer some of the most fundamental questions about humans. Also, it examines everything from the blue whale breaching the ocean's surface for air to the sugar ant crawling along a kitchen counter to the bacteria that runs the digestive process.

The life sciences or biological sciences encompass the branches of science that basically involve the scientific study of life and organisms – such as microorganisms, plants, and animals, including human beings, too. In addition, life science is regarded as one of the two main branches of natural science, the other being physical science, which is mainly concerned with non-living matter.

Further, life sciences focus on a particular kind of organism. For an instance, zoology is the field of studying animals, while on the other hand, botany is the field of studying plants. Some other field of life sciences focus on certain aspects that are common for all or many life forms, such an genetics, anatomy, etc.

Some focus on the micro-scale (i.e., molecular biology, biochemistry, etc.) while, others on large scales (i.e., cytology immunology, ethology, ecology, etc.). Another significant branch of life sciences involves the understanding the mind's neuroscience.

Life science may be basic or applied. The key aim of basic life science is to acquire knowledge and deep understanding of the natural world. Whereas, the aim of applied science is to find out solutions to practical issues and problems. Discoveries in the field of life sciences are effective and helps in improving the quality as well as standard of life. Also, it has numerous applications in sector of health, medicine, agriculture, pharmaceutical and food science industries.

This book will introduce the readers to the field of life science and the fundamentals of it. It is precisely designed for readers or students with no prior experience related to the field of life science, and touches upon a diversity of fundamental topics. By the end of the book, readers will understand the basic knowledge of life science and the key concepts involved.

A

A Site

During translation, one of the three binding sites for tRNA on a ribosome. The tRNA containing the next amino acid to be introduced to the polypeptide chain is kept in the location.

ABC Hypothesis

A flower development model that specifies three classes of organ identity genes that control the production of the four types of floral organs.

Abiotic

Refers to an environment's physical and chemical qualities

Abortion

The termination of an ongoing pregnancy.

Abortive Transduction

The transmission of bacterial genes into a new host by a viral vector without the new genetic material being integrated into the recipient cell's genome. The transduced genes, though, may survive in the cell as a plasmid for a period of time.

Abscisic Acid (ABA)

a plant hormone that inhibits growth and frequently counteracts the effects of growth hormones. It has a number of actions, including increasing seed dormancy and improving drought resistance.

Absorbance

A measure of how much light is received when it passes through a coloured liquid or solid.

Absorptiometer

A device for determining the absorption of a sample, often a coloured liquid. It is made up of a light source, a sample cell, and a photomultiplier, or any other light-sensitive cell, all of which are connected to an appropriate meter or recorder.

Absorption

The uptake of tiny nutrient molecules by an organism's body, which is the third phase of food processing in animals.

Absorption Spectrum

The image that is formed when electromagnetic radiation from a source exhibiting a continuous spectrum passes through a material. Bands will emerge in the same place as lines in the material's characteristic emission spectra if the material is in the gaseous phase. These bands are widened in solids or liquids and are often used to identify or measure the item.

Abyssal Zone

The section of the ocean's benthic zone that is between 2,000 and 6,000 meters deep

Acanthodian

Any of a genus of jawed aquatic animals that lived throughout the Silurian and Devonian periods.

Acceptor Site

An enzyme's active site or an area on an organism, tissue, or cell that a biological substance identifies and has a reaction towards, in a very precise manner.

Accessory Fruit

A fruit or group of fruits where the fleshy components are obtained in part or fully from tissues apart from the ovary.

Figure 1. Pseudocarps.

Source : Image By Wikimedia Commons

Acclimatization

Physiological response of an organism to a modification in a component to adjust in modified environment.

Acellular

A term used to describe an organism or tissue that is made up of a mass of protoplasm that isn't split into cells. The multinucleate hyphae in fungus are a good example.

Acentric

A term used for a chromosome or chromosome fragment that lacks a centromere.

Acer Pseudoplatanus

Refers to the sycamore tree, which is a root cause of heterotrophic and autotrophic plant cell suspension cultures that have been utilized extensively in research.

Acesulpham K

A saccharin-like synthetic sweetener. It has 150 times the sweetness of sucrose.

Acetaldehyde

A component in the glycolysis process that leads to the production of ethanol. It is made when pyruvate is decarboxylated in a reaction mediated by pyruvate decarboxylase.

Acetic Acid

An organic acid that is produced as a result of fermentation process. It is best known as the active ingredient in vinegar.

Acetic Acid Bacteria

Bacteria belonging to the genus Acetobacter or Acetomonas that oxidizes primary alcohols and aldehydes to the acids they produce.

Acetic Anhydride

An acid anhydride which is produced as a result of a dehydration reaction between two acetic acid molecules.

Acetobacter

Such bacteria are gram-negative rods with an aerobic metabolism that are utilized to make acetic acid or vinegar. They are also implicated in the contamination of beer and the deterioration of wine. A genus of the eubacteria.

Acetoclastic Bacteria

Organisms that produce methane solely from acetic acid during anaerobic digestion. They grow gradually over time, with doubling rates of several days, and are frequently the rate-limiting phase in biogas generation.

Acetogenic Bacteria

Organisms that convert fatty acids to acetic acid, particularly propionic and butyric acids. They can only evolve when there is no hydrogen present.

Acetone

An essential commercial solvent produced by fermentation in some bacteria, most notably *Clostridium acetobutylicum*.

Acetone/Butanol Fermentation

An industrial method that produces a combination of acetone and butanol by fermenting sugars (for example, molasses) using *Clostridium acetobutylicum*. This was the first large-scale fermentation employing only pure culture and aseptic techniques.

Fermentation was once the primary source of these solvents. With the development of the petrochemical sector, however, practically all factories were shut down. This technique is gaining popularity because it might be used to make cosolvents for use in lead-free petroleum blends.

Acetone Powder

A protein preparation that results from the isolation of enzymes. The frozen sample is homogenized in acetone at temperatures below - 30°C to avoid autolysis and/or inactivation of enzymes by inhibitory compounds.

Acetyl-DL-Amino Acid

The acetylated state of an amino acid that is generated by acetic anhydride treatment of hydrolyzed protein.

Acetylcholine

Refers to the acetyl ester of choline, which works as a neurotransmitter in the central and peripheral nervous systems of vertebrates and invertebrates. It does so by attaching to receptors and modifying the permeability of the postsynaptic membrane to particular ions, either depolarizing or hyperpolarizing the membrane.

Acetyl-Coenzyme A (Acetyl-COA)

A compound containing an acetyl group and coenzyme A formed during the oxidative decarboxylation of pyruvate and the oxidation of fatty acids.

It serves as a connection between glycolysis and respiration's TCA cycle, is a glyoxylate cycle intermediate, and is essential in the production of a wide range of primary and secondary metabolites, including fatty acids, certain amino acids, and terpenoids.

Acetylene Reduction Test

A precise method for determining an organism's ability to fix nitrogen. It is premised on the fact that the enzyme system that reduces nitrogen gas to ethylene (dinitrogen) can also reduce acetylene to ethylene.

In a closed flask, the material being study is incubated in an acetylene atmosphere. Periodically, the gas phase is sampled, and the component gases are separated using gas chromatography.

Acetylmuramic Acid

A type of amino sugar made from o-glucosamine and lactic acid. It is present in bacterial cell walls as part of the mucopeptides.

A Chromosome

One of the chromosomes that constitute a nucleus of a eukaryotic cell's normal complement.

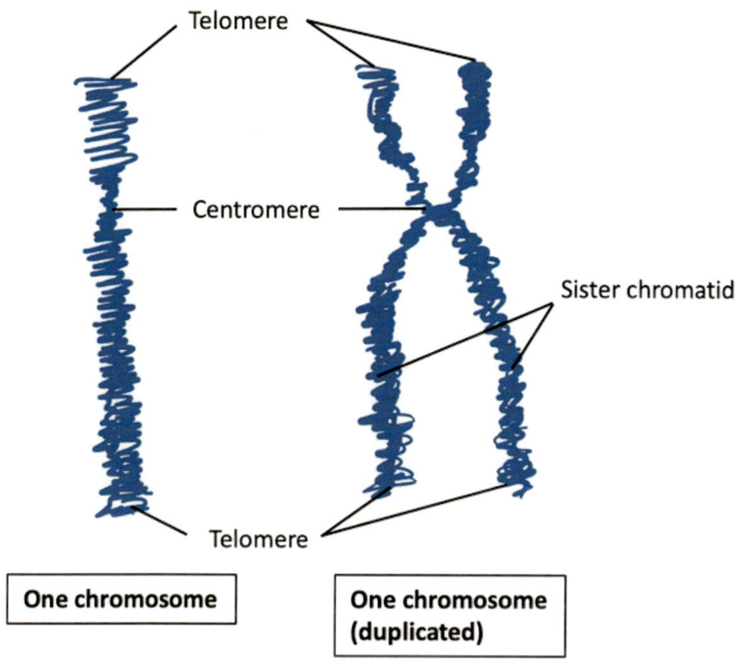

Figure 2. Schematic Diagram of Chromosome Structure.

Source : Image By Wikimedia Commons

Acid

A chemical that raises the concentration of hydrogen ions in a solution.

Acid-Base Balance

The acid-base ratio in the blood. Bicarbonate and carbonic acid are found in blood in a 20:1 ratio, which keeps the blood pH at 7.4. If this ratio changes owing to carbonic acid build up, the kidney's bicarbonate reabsorption is modified to balance and restore the proper ratio.

Acidification

A procedure in which the pH is lowered or acid is introduced to lower it. Acidification after shock loading might cause issues in anaerobic digestion. Acid-producing bacteria can grow and metabolize at a quicker pace than methanogenic bacteria, which are less resistant to acid environments.

As a result, if acid production rates increase for whatever cause, the imbalance is exacerbated when acid manufacturers proliferate quicker and acid-consuming bacteria are suppressed. This might result in the biogas-producing system collapsing.

Acidophilous Milk

A substance derived from the fermentation of milk using *Lactobacillus acidophilus*.

Acid Precipitation

Rain, snow, or fog containing a pH higher than 5.2.

Acid Protease

A protein-hydrolyzing enzyme with maximal activity and stability between pH 2.0 and 5.0. Acid proteases have a low isoelectric point and a low amount of basic amino acids. Sulphydryl reagents, heavy metals, and metal chelators have little effect on them, although pH levels above 6.0 neutralize them.

They are common in the food and beverage industry. Commercially important acid proteases are made from fungal sources and are divided into two categories: pepsin-like and rennin-like enzymes. Aspergillus species generate pepsin-like proteases, while Mucor species produce rennin-like proteases.

Acoelomate

A solid-bodied animal, metazoan, without a cavity between the stomach and the exterior body wall, such as nematodes (threadworms).

Aconitase

An enzyme that catalyzes the transformation of citric acid to iso-citric acid through aconitic acid in the TCA cycle.

Aconitic Acid

A TCA cycle intermediate that normally only arises as an enzyme-bound intermediate in the presence of aconitase. Mutant Aspergillus strains, on the other hand, can generate significant amounts of this chemical.

Aconitine

A extremely poisonous diterpenoid alkaloid made by combining Aconitum and Delphinium.

Aconta

Algae whose life cycle does not include a flagellate stage. This is typical of red algae (Rhodophyta), which provide a significant amount of industrial polysaccharides, including agar.

Acquired Immune Deficiency Syndrome (AIDS)

An illness caused by a virus that has a high death rate. AIDS was initially used to describe a tightly circumscribed disease characterized by Kaposi's sarcoma and Pneumono-cystis carinii pneumonia symptoms. It is now recognized, however, that it also includes a number of other life-threatening acute diseases.

AIDS refers to a set of illnesses caused by a deficiency in cell-mediated immunity caused by a single, recently identified virus. Human T-cell lymphotrophic virus type III, lymphadenopathy-associated virus, and AIDS retrovirus are all names for this virus. The virus looks to be a retrovirus from the Lentivirinae subfamily, which has been discovered in sheep, horses, and goats in the past.

The AIDS virus causes a considerably broader range of diseases than was first thought, notably progressive encephalopathy (i.e. degeneration of the brain). The virus penetrates a limited number of mature T-helper lymphocytes and brain cells, where it multiplies slowly in both an integrated and unintegrated form.

Infectious virions are produced by both types of cells. The integrated virus gets embedded in the DNA of brain cells, making treatment difficult. The virus multiplies in T-helper cells and kills them, resulting in mild or severe immunological insufficiency depending on the number of cells killed.

The illness is transferred by blood transfusions and blood clotting factor preparations, as well as human interaction, because an affected individual's blood and plasma are extremely contagious. Antibodies are not known to harm the virus when they are produced, and the virus demonstrates antigenic drift.

The ENV gene, which codes for the envelope glycoprotein, varies greatly amongst viral isolates, making vaccine development difficult.

Acridine

A mutagen that induces base pair additions and deletions, particularly in plasmids and other non-chromosomal DNA.

Acridine Orange

Dimethyl-amino acridine hydrochloride, which is a nucleic acid staining dye. When intercalated into double-stranded DNA, it fluoresces at 530 nm, and when ionically attached to the phosphate backbone of single-stranded DNA or RNA, it fluoresces at 640 nm.

Acrocentric

Term for a chromosome having the centromere at one end.

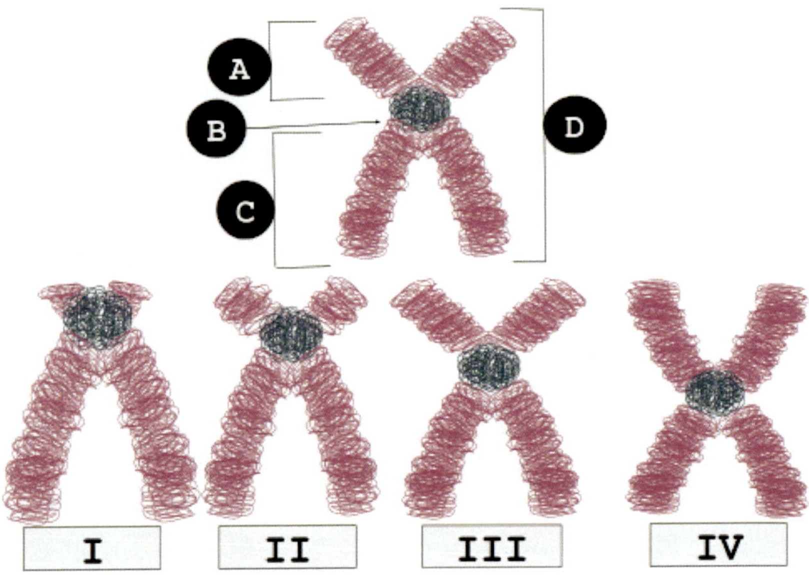

Figure 3. Classifications Of Chromosomes.

Source : Image By Wikimedia Commons

Acrosomal Reaction

The release of hydrolytic enzymes from the acrosome, a vesicle in the tip of a sperm, as the sperm approaches or meets an egg.

Acrosome

A vesicle found at the sperm's tip that contains hydrolytic enzymes and other proteins that aid in the sperm's journey to the egg.

Acrylamide

The monomer used to make polyacrylamide gels, which are used to separate proteins and nucleic acids via electrophoretic separation.

Actin

A globular protein that forms microfilaments (actin filaments) in muscle and other cells by linking into chains, two of which twist helically around each other. Muscle and flagella contain this protein, which is present in contractile systems.

Actinomycetes

A class of gram-positive eubacteria that are rod-shaped, filamentous, and do not have cross-walls, remotely similar to microscopic mushrooms (diameter smaller than 0.5 -tm).

This group comprises oxidatively metabolic soil organisms (such as Streptomyces, Nocardia, and Mycobacterium), along with parasitic or fermentative forms, such as the species Actinomyces proper. Antibiotics and enzymes are manufactured using Streptomyces species.

Actinomycin D

A Streptomyces-derived antibiotic that inhibits the production of rRNA.

Action Potential

An electrical signal that proliferates (travels) along the membrane of a neuron and other excitable cell as a nongraded (all or none) depolarization. During the passage of an impulse over the membrane of a nerve or muscle cell, the potential across the membrane increases from roughly -70 m V to around + 30 m V.

Action Spectrum

A plot depicting the strength or size of a biological reaction by expressing it as a function of electromagnetic radiation wavelength. The visible and/

or ultraviolet areas are commonly depicted for action spectra. For a photo-mediated or light-dependent process, the size of the response will match the absorption spectrum of the molecule involved.

As a result, the action spectrum may be utilized to establish a causal link between a pigment or photoreceptor and a biological reaction.

Activated Sludge

A combination of aerobic organisms (biomass) formed during the sewage or wastewater treatment process with activated sludge.

Activated Sludge Process

The process of increasing contact with high concentrations of rapidly multiplying microorganisms in the vicinity of sufficient dissolved oxygen in sewage and wastewater treatment whereby the pace at which particles are washed out of the system balances the growth of new biomass.

Agitation or air injection are used to maintain the dissolved oxygen level. To maximize contact time and sustain a high degree of biological activity, the volumetric flow rate of waste sludge is maintained relatively stable, and settled solids are cycled back to the early aeration stage.

Activated Support

A matrix material that has been processed to produce a significant number of chemically reactive groups on the surface.

Activation Energy

The quantity of energy that should be absorbed by reactants before a chemical reaction can begin. This is also known as activation free energy. The amount of energy required to create a transition state complex. This may be shown by looking at the reaction.

$$A+B=C+D$$

Only when molecules A and B form a transition complex AB with potential energy larger than A + B will a reaction occur. As the reaction must be exogenic, this transition complex can disintegrate to generate the products C + D, which have a lower potential energy than A + B.

The rate of reaction is relative to the quantity of AB in the solution. At a particular temperature, the activation energy is equal to the amount of energy necessary to move all of the molecules in one mole of A and B to the upper end of the energy barrier distinguishing A and B from C and D. Through reduction in the activation energy, enzymes speed up the process.

Activator

A substance that is required for the activity of a specific enzyme but it does not serve as a substrate or add value to the final product. An activator can operate on an enzyme by attaching to a particular allosteric site.

Active Biomass

The percentage of a cell culture that is active microbiologically.

Active Immunity

Immunity resulting from the body's own generation of antibodies instead of the administration of antibodies via injection, for instance. Active immunity, which can result in the development of the disease or through vaccination, provides a high level of resistance to recurrent infection, by using killed or attenuated vaccines or toxoids.

Immunity develops over few days to a few weeks, lasts a long time (potentially for life), and is easily reactivated by a booster injection of the needed antigen. As a preventative measure or prophylactic, active vaccination is utilized.

Active Site

The enzyme region that binds the substrate to a product during catalytic conversion . The fundamental hypothesis of enzyme activity proposes that binding at the active site results in the creation of an enzyme-substrate complex. The activation energy is lowered by electrostatic attraction, repulsion, and other intermolecular physical factors, making intermolecular conversions easier.

Active Transport

A biological process that causes molecules to pass through a membrane against a concentration gradient. Active transport is coupled with a particular carrier protein or lipoprotein molecule and necessitates metabolic energy (typically A TP or an electrochemical gradient).

Acylamino Acid

An amino acid that has had an acyl group added to it.

Adaptation

The adaptation of an organism to its surroundings. It refers to the methods through which a microbial culture adjusts to grow on a certain type of media.

Adaptive Enzyme

During adaptation, an enzyme that is produced in response to an external stimulus. This term has been superseded by inducible enzyme, which is currently widely used.

Adaptive Value

A comparison of an organism's (or genotype's) reproductive efficiency to that of other species (or genotypes). Selective value is another name for it.

Adaptive Immunity

A kind of vertebrate defense regulated by b lymphocytes (b cells) and t lymphocytes (t cells). It demonstrates uniqueness, memory, and self-nonself recognition. Acquired immunity is another name for it.

Adaptive Radiation

A phase of evolutionary transition during which groups of animals develop numerous new species, each with its own adaptations that enable them to perform distinct ecological functions in their environment.

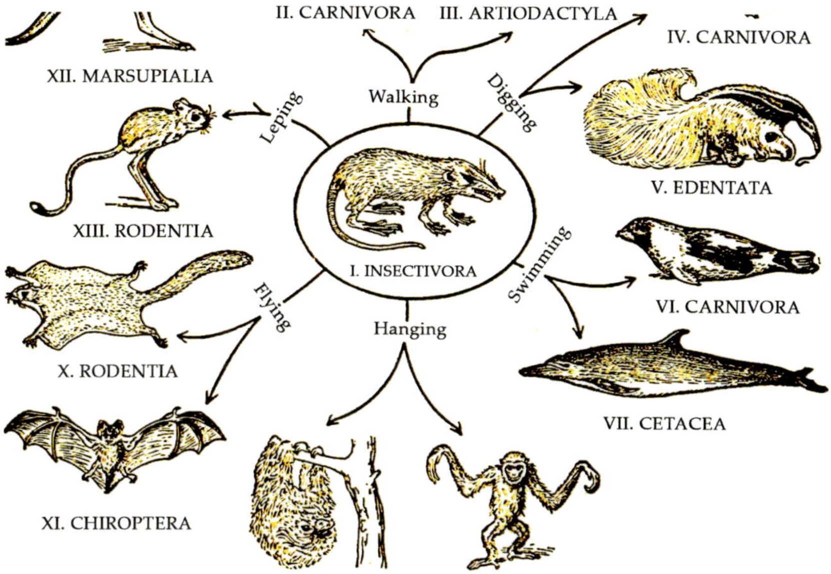

Figure 4. Image Showing Adaptive Radiation.

Source : Image By Flickr

Adaxial

A term for the surface of a leaf or other lateral organ that is nearest to the top of the axis on which it is carried

Addition Rule

The likelihood of any one of two or more mutually exclusive events happening can be calculated by summing their individual probabilities, according to the law of probability.

Additive Genes

The genes that codes for same character and phenotype is the result of their combined effect. The alleles of gene are neither dominant or recessive.

Additive Recombination

A type of gene recombination that happens when a new sequence of DNA is inserted into an existing genome without causing DNA loss in the host genome.

Adenine (A)

A purine base that is found in nucleic acids and coenzymes such as NAD and FAD; it is also found in adenosine.

Adenosine

A nucleoside constituted of adenine connected to o-ribose through a glycosidic bond. The primary molecules involved in energy transmission in biological systems are phosphorylated forms of adenosine (AMP, ADP, and ATP).

Adenosine Diphosphate (ADP)

A nucleotide comprising adenine, o-ribose and two phosphate groups. In biological energy metabolism, ADP plays a key role.

Adenosine Monophosphate (AMP)

A nucleotide consisting of adenine and o-ribose and one phosphate group. The production of ADP or ATP is caused by the addition of one or two more phosphate groups.

Adenosine Triphosphate (ATP)

A nucleotide made up of adenine, o-ribose and three phosphate groups. Pyrophosphate bonds connect two of the phosphates, resulting in a substantial shift in free energy upon hydrolysis.

Phosphorylation processes in respiration and photosynthesis create ATP, which is essential in cellular energy metabolism. By linked processes, the energy stored in ATP is utilised to generate new molecules.

Adenovirus

A group of naked icosahedral viruses containing double-stranded DNA and up to 12 distinct kinds of protein molecules. In humans, adenoviruses lead to acute respiratory infections.

Adenylyl Cyclase

This enzyme transforms ATP into cyclic amp as a reaction to an extracellular stimulus.

Adhesion

The adhering of one material to another via hydrogen bonds, such as water to plant cell walls.

Adipic Acid

A six-carbon dicarboxylic acid that is employed in the manufacturing of nylon.

Adipose Tissue

Fatty tissue that is made up of clustered cells that contain a lot of fat. A single enormous fat droplet frequently displaces the contents of mammalian cells, notably in subcutaneous cells, the mesentery, and the mediastinum. Adipose tissue serves as a source of energy, as well as protection for vulnerable vital organs and thermal insulation.

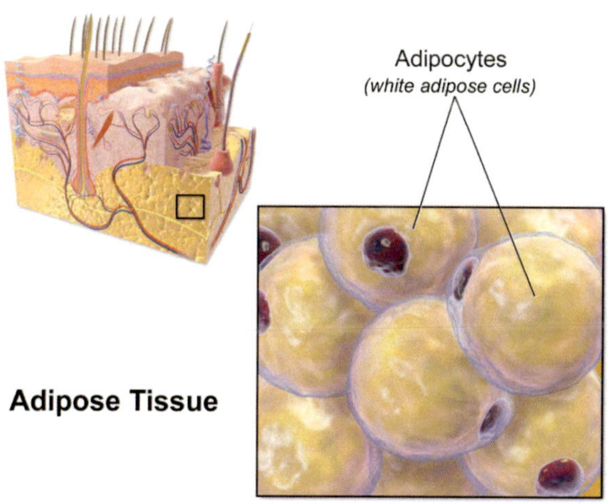

Figure 5. Adipose Tissue.

Source : Image By Wikimedia Commons

Adjuvant

A chemical that isn't antigenic but boosts antibody production when combined with an antigen. Adjuvants are utilized therapeutically because they aid in the generation of antibodies against tiny quantities of antigen and extend the time in which antibodies are produced. Adjuvants function by triggering an inflammatory response that results in an influx of antibody-forming cells in a specific area.

Adrenal Gland

In animals, a pair of complex endocrine glands that are located on the front or upper surface of each kidney. The adrenals are separated into two regions: an outer cortex that originates from the embryo as lateral mesoderm and an inner medulla that originated from the neural crest.

Corticosteroids are a group of steroid hormones secreted by the cortex. Adrenaline and noradrenaline are two catecholamines produced by the medulla.

Adrenaline

A catecholamine hormone released by the mammalian adrenal medulla, commonly known as epinephrine. During stressful situations, the sympathetic nervous system stimulates its secretion. Adrenaline boosts blood glucose levels and promotes blood flow to skeletal muscle.

Adrenergic

A term for a nerve fibre or nerve terminal that produces the neurotransmitters adrenaline or noradrenaline.

Adrenocorticotropic Hormone (ACTH)

A polypeptide hormone with 39 amino acids that is naturally produced by the pituitary gland's pars distalis in mammals. ACTH activates the adrenal cortex to produce corticosteroid hormones and pigment cells to produce melanin.

The production is controlled by corticotrophin-releasing factor, a hypothalamic hormone, and vasopressin, and also feedback processes wherein corticosteroids affect pituitary and hypothalamic activity.

Adsorption

The adherence of molecules to solids in physical chemistry. Adsorption chromatography, which is the process employed to purify proteins and other macromolecules, plus various methods of cell or enzyme immobilisation, are based on this process.

Adventitious

A term for a structure that is not inherent and has developed in an unusual locationunusual manner from unexpected place.

Aeration

(1) The addition of oxygen (or air) to aerobic fermentation events. (2) The injection of any gas into a liquid, – for example carbon dioxide.

Aerobe

A living entity that needs oxygen to breathe and function.

Aerobic Reactor

A fermenter or bioreactor equipped with an aeration system for the cultivation of aerobic organisms.

Aerobic Respiration

The entire process of thoroughly oxidising carbohydrates to carbon dioxide and water through molecular oxygen. Glucose phosphate is metabolised to pyruvate in the first step (glycolysis), which is standard for both aerobic and anaerobic respiration.

Pyruvate is subsequently transformed to acetyl-CoA, which is then combined with oxaloacetic acid to generate citric acid, which enters the TCA cycle. In a cyclic mechanism, citric acid is metabolised to reconstruct the acceptor molecule oxaloacetic acid.

This leads to the release of two molecules of carbon dioxide, as well as the passage of hydrogen atoms and their electron equivalents to an intermediate electron transport chain, which culminates in the creation of water from molecular oxygen.

Energy is preserved in the form of ATP while progression through the series of redox couples which make up the intermediary electron transport chain.

Aerobic Waste Treatment

A biological technique that involves the growth of aerobic bacteria on wastewater to reduce the biochemical oxygen requirement.

Carbon molecules are used by rapidly developing organisms as a supply of material for cell expansion and metabolic energy, resulting in cell biomass and carbon dioxide. See also activated sludge process, algal oxidation pond, deep shaft system, and trickling filter.

Aerogel

A hard, prepared matrix with pores that have been filled with a solvent. The matrix will create aerogels in any solvent since it is not a true gel. Porous glass beads and titanium spherules are two examples. Aerogels are utilised in chromatography for gel filtration, as ion exchanger carriers, and xerogel materials.

Aerosol

(1) A gaseous system comprised of tiny droplets or colloidal particles. (2) Infectious organisms suspended in a tiny suspension of water droplets.

Afferent

A term for nerve or neurone that carries information from peripheral receptors to the central nervous system

Affinity

An inherent orientation towards a substance or item. The word refers to the unique contact that happens between an enzyme and its substrate, or an antigen and an antibody, which may be employed in immunoassay, affinity labelling, and affinity chromatography procedures.

Affinity Chromatography

A chromatographic approach that relies on a molecule's unique affinity for another. For example, enzymes can be separated by attaching an analogue of their typical substrate to an inert matrix. The needed enzyme will be preserved or slowed if a solution of mixed proteins is passed through a column filled with such a matrix because of its affinity for the bound substrate.

After that, the protein is recovered by eluting the column with a viable alternative that has a pH or ionic concentration that reduces the affinity. Antibodies (particularly monoclonal antibodies) are purified using this method by passing them through a column filled with a comparable matrix to which an appropriate antigen has been complexed.

Affinity-Isolated Antibody

An antibody that has been highly purified using an affinity column comprising a matrix to which the antigen for the necessary product has been linked.

Affinity Labelling

A non-radioactive method for labelling nucleic acids to be used as probes. The probe is connected to a group that does not interfere with the probe's capacity to hybridise with complimentary sequences. The affinity-labelled duplex is then reacted with a signal molecule that has a binding site for the affinity label after hybridization.

Fluorescent antibodies, colour-changing enzymes, and chemiluminescent catalysts are examples of signal molecules. Biotin, for example, can be chemically bonded to the probe DNA nucleotides. The protein avidin binds to biotin and can be labelled with a fluorescent marker. The method's response can be improved by adding lengths of biotinylated nucleotides to the probes.

Affinity Precipitation

A system involving the creation of a precipitate as a result of two types of molecules reacting at a specific contact point (e.g., enzyme and substrate, antibody and antigen).

Aflatoxins

A class of extremely poisonous and typically carcinogenic compounds generated by the molds *Aspergillus flavus* and *Aspergillus parasiticus* all through their development on food or animal feed. Consumption of even modest levels of these poisons (the LD50 could be less than 0.35 mg/kg body weight) can be harmful to one's health.

Toxigenic Aspergillus species are extensively spread in nature, and the risk of contamination is significant, particularly in grains and oilseed residues that have not been properly dried or kept in damp settings. Adulterated materials fed to cows might result in contaminated milk.

There are almost a dozen different types of aflatoxin. Aflatoxins are replaced coumarins chemically, with B1 being the most poisonous and carcinogenic.

Agammaglobulinaemia

A condition in which people are born without enough gamma-globulin and produce little or no antibodies. As a result, such individuals are more vulnerable to bacterial infections.

Agamospermy

An apomixis that isn't vegetative propagation.

Agar

A complex polysaccharide generated by red algae, particularly Gelidium and Gracilaria species. Agarose and agaropectin are polysaccharides found in it. Agar is also used in the food industry and as a microorganism culture matrix. It is especially well-suited for the latter, as just a few bacteria develop enzymes sufficient for liquefying the gel.

Agaropectin

A polymer that is similar to agarose but has D-glucuronic acid as well as minor quantities of other sugars, such as sulphate esters.

Agarose

A polysaccharide gum made from seaweed that contains modest quantities of D-xylose and alternating (1,3)-linked D-galactose and (1,4)-linked 3,6-anhydro-L-galactose residues. At C-6, several D-galactose units are methylated. In chromatography and electrophoresis, agarose is utilised as a gel medium. It has a wide molecular fractionation range that includes viruses and microparticles.

Agarose Gel Electrophoresis

A separation process in which the stationary phase is an agarose gel. As agarose gel electrophoresis can separate DNA molecules based on their molecular weights, it is useful in gene modification and sequencing. Ethidium bromide is used to identify the bands on the gel, allowing amounts as low as 0.5 mg DNA to be identified under UV light.

Figure 6. Loading A DNA Sample into An Agarose Gel For Electrophoresis.

Source : Image By Wikimedia Commons

Age Structure

The proportion of people of each age group in a population.

Agglutination

The interaction of antigens on the cell surface with antibodies to construct bridges joining the antigen determinants, thereby resulting in the development of clumps or floes of germs or cells, such as erythrocytes. Blood groups, microorganisms, animal cell culture lines, and other things are all detected via agglutination reactions.

Agglutinogen

An antigen found in a bacteria that, upon administration/injection into an animal, stimulates the formation of agglutinin

Aggregate Fruit

A fruit produced by a single flower with many carpels.

Aggregation

The clumping or flocculation of single organisms or cells to create groups.

Aging

The physiological and genetic changes that cause a cell or organism to age and eventually die.

Agitation

Shaking or other violent irregular movements or motions that are used to combine ingredients.

Aglycone

In the production of glycosides, a molecule with a nucleophilic atom (typically oxygen found in an alcohol, phenol, or organic acid) that substitutes the hydroxyl group on the anomeric carbon atom (C-1 of aldoses or C-2 of ketoses). S- and N-glycosides are produced as well.

Agranulocyte

White blood cell with no cytoplasmic granules. Lymphocytes and monocytes are examples of such cells.

Agricultural Alcohol

Ethanol created by the fermentation of agriculturally sourced raw materials. Sugar cane, maize, grapes, molasses, potatoes, damaged fruit, wheat, cassava, and rice are the most common agricultural alcohol sources.

Agrobacterium

A genus of bacteria that contains *A. tumefaciens*, a bacterium that causes crown gall tumours in gymnosperms and dicotyledonous plants. A DNA plasmid induces the sickness (tumour-inducing or Ti plasmid). In the total absence of the bacteria, this is able of independent replication within the cells of the plant host.

The shape of the tumours formed as a result of infection differs depending on the bacterial strain and plant species engaged. The exact morphology is determined by how much the host plant's hormone system has been affected. Undifferentiated tumours generate viable teratomatic plants, whereas others do not.

The tumor's plant cells have two novel characteristics - first, their development is independent of phytohormones or other plant growth factors, and second, they carry one or more of the uncommon guanido acids called opines. Octopine and nopaline are the most prevalent opiates.

Plants regenerated from nopaline-containing tissue keep synthesizing nopaline, and crown gall tumours continue to synthesise opianes in tissue culture. The capacity of tumour cells to synthesis opianes is a characteristic imparted to them by bacteria that can use opianes as their only supply of nitrogen.

The bacteria do not enter the plant cells that have been transformed to tumour cells; instead, they enter the intercellular gaps and cling to the healthy plant cell walls. The genetic information defining bacterial utilisation of opines and their synthesis by plants is incurred on the Ti plasmids.

These plasm ids are commonly recognised as ideal for use as DNA vectors in plant gene modification since they can replicate in plant cells. See also plant cloning vehicle.

AIDS (Acquired Immunodeficiency Syndrome)

Symptoms and indicators that arise during the late stages of HIV infection, as indicated by a decrease in the number of t cells and the advent of certain secondary infections.

Air Filter

A device for eliminating tiny particles from the air. Air filters are used to keep infectious or contaminated particles out of a space (such as a clean room) or a bioreactor, as well as to keep pathogens or genetically modified organisms from escaping into a ventilation stream from a containment facility or fermenter.

Air Lift Fermentation

A procedure in which air is introduced at the base of a central draught tube to keep the cell suspension mixed and aerated. This causes liquid to flow vertically. In the industrial synthesis of single cell protein, large-scale air lift fermenters are employed.

Air Monitoring

The process of determining the quality of an air stream. It's utilised to identify biological and non-biological contaminants in order to avoid contamination and/or maintain asepsis. Air quality may be measured by accumulation of particles on ultrafilters and subjecting them to nutrients, allowing imprisoned cells to be identified and counted based on the colonies produced.

Air Sterilization

The process of removing contaminated organisms from the air before it is used to supply oxygen or to mix cultures. To eliminate organisms with dimensions as tiny as 0.5-tm from extremely large quantities of air, air sterilisation methods are necessary. Although the heat generated during air compression can help kill some organisms, it is usually vital to filter the air with fibrous, sintered, or acetylated polyvinyl alcohol sheet filters.

Alanin

The amino acid alanine in protein sequences and elsewhere is abbreviated as ALA. Alanine is made by direct fermentation with bacteria such as Corynebacterium, Brevibacterium, Arthrobacter, Microbacterium, or Pseudomonas with a medium based on glucose or molasses. Pseudomonas or Xanthomonas cultures with significant L-aspartic-B-decarboxylase activity may also generate alanine from aspartic acid.

Albumen

The water-soluble protein found in egg white. It is made up of a variety of conjugated proteins, including ovalbumin, which has a carbohydrate prosthetic group.

Figure 7. Albumen – Fried Egg.

Source : Image By Pixabay

Albumin

A simple protein group that is soluble in water and coagulates when heated. Albumins in the blood are called serum albumins.

Alcaligenes

A genus of eubacteria. Gram-negative rods, some of which are facultative anaerobes, make up the bacteria. These organisms can cause milk to deteriorate, producing slimy or ropy milk.

Alcohol

A chemical molecule with the generic formula ROH, in which R stands for an alkyl group.

Methanol is the lowest molecular weight alcohol. Alcohols, such as acetic acid, ethyl acetate, ethyl ether, and polythene, are commonly utilised as industrial solvents and as precursors for the production of different acids, esters, aldehydes, and polymers. Alcohols can be utilised as fuels, octane boosters, or cosolvents in petroleum blends in internal combustion engines.

Alcohol Dehydrogenase

The enzyme that catalyzes the acetaldehyde-to-ethanol conversion.

Alcohol Fermentation

An anaerobic metabolic process found in yeasts, bacteria, and fungus. Glycolysis converts glucose to pyruvic acid, which results in the generation of carbon dioxide and (alcohol) ethanol. This is the most common method for making drinkable, agricultural, and fuel alcohols.

Large-scale fuel alcohol production (e.g., in Brazil, over 11 billion litres of fuel alcohol are generated each year from sugar cane), and the production of beers, wines, and spirits, all rely on yeast-based alcoholic fermentations.

Alcohol Production

Ethanol is mostly produced by fermentation involving yeasts that are cultivated on sugar-containing substrates, which can be directly produced juices or derived from starch hydrolysis.

The production of alcoholic beverages for direct consumption (beers or wines), alcoholic beverages for consumption after distillation to significantly boost the alcohol content (spirits such as gin, whisky, vodka, brandy, etc.), or the production of alcohol to be used as a chemical feedstock, fuel additive, or fuel in its own right can all be classified as alcohol production.

Aldehyde

A chemical with the generic formula RCHO, where R represents an alkyl group. When aldehydes are reduced, they produce alcohols, and when they are oxidised, they produce acids.

Aldolase

The enzyme that catalyzes the reversible conversion of fructose bisphosphate into one molecule of glyceraldehye phosphate and another molecule of dihydroxyacetone phosphate. It is a glycolysis enzyme.

Aldonic Acid

A chemical formed when an aldose sugar is oxidised at the C-1 site.

Aldose

An aldehyde-containing monosaccharide.

Aldosterone

A steroid hormone that regulates the transit of sodium ions ($Na+$) and potassium ions ($k+$) in kidney tubules.

Alduronic Acid

A chemical formed by oxidising an aldose at the C-6 position.

Ale

A top-fermented beer, which is one made using the Saccharomyces cerevisiae yeast, which rises to the top of the fermentation and is skimmed off. There are several sorts of ales, each of which is defined by its colour and body. Pale ale is characterised by its light colour and intense hop bitterness.

Brown ale (mild ale) is darker and sweeter than pale ale. Stout is a full-bodied, black ale with a low amount of hop bitterness.

Algae

Aquatic plants that do not have genuine roots, stalks, or leaves yet possess chlorophyll a and other distinguishing colours. The eukaryotic Rhodophyta, Chromophyta, Chrysophyta, Phaeophyta, Euglenophyta, and Chlorophyta are the primary categories. The prokaryotes Cyanophyta are classified as blue-green algae in several classification systems.

Several taxonomic systems, on the other hand, classify them as bacteria or cyanobacteria. Overgrowth of algae in contaminated water can pose major difficulties. Gums are made from a variety of algae that may be found in nature (such as agar, agarose, alginate and carrageenan). Microbial culture, food products, chromatography, and electrophoresis all use these gums.

Algae can be farmed for fish feed, pigments, lipids, and other compounds (including hydrocarbons), as well as single cell protein. In the cultivation of paddy rice, certain nitrogen-fixing blue-green algae are produced as organic fertilisers.

Figure 8. Green Algae in Algonquin Provincial Park, Pen Lake, Ontario, Canada.

Source : Image By Wikimedia Commons

Algal Biomass

Algae cultures like Chlorella, Spirulina, and Scenedesmus that are cultivated as animal feed or food.

Alginate

An acetylated polymer comprising mannuronic acid and guluronic acid, which is an anionic polysaccharide. Brown seaweed is used to extract it for commercial purposes. *Azotobacter vinelandii* and *Pseudomonas aeruginosa* are two species that create microbial alginates.

In the presence of calcium ions, alginates create viscous gels. Commercially, many viscosity grades are offered. These have different guluronic acid to mannuronic acid ratios. Alginates are commonly utilised in food items, textile printing, and microbial immobilisation.

Algorithm

A series of steps defining a procedure or formula for solving a problem that can be coded into a programming language and executed. Bioinformatics algorithms typically are used to process, store, analyze, visualize, and make predictions from biological data.

Alimentary Canal

In animals, the route from the mouth to the anus via which foodstuffs travel to be broken down and assimilated into the body. For accommodating a wide range of foodstuffs, the alimentary canal has a vast variation in shape and biochemical specialisation.

Aliphatic

A term for open-chain organic compounds – for example, paraffins and olefins

Aliquot

A tiny sample of a material in a known volume solution.

Alkaline Hydrolysis

A type of chemical hydrolysis that uses an alkali, such as sodium hydroxide.

Alkaloid

A nitrogen-based substance. Purine, pyrimidine, quinoline, iso-quinoline, pyridine, reduced pyridine, and indole are alkaloids with different chemical structures and are classed according to the kind of heterocyclic group they contain. Plants generate the majority of alkaloids, and most of them have powerful pharmacological properties.

Analgesics (morphine), stimulants (caffeine, nicotine), muscle relaxants (strychnine), opiates (opium), vasoconstrictors (scopolamine), anaesthetics (cocaine), tranquillizers (reserpine), and psychedelic drugs are all examples of alkaloids (psilocybin).

Alkanes

Components of oil that are saturated aliphatic hydrocarbons. A number of microorganisms rapidly metabolise normal (straight chain) n-alkanes with more than nine carbon atoms to create C2 chemicals via beta-oxidation. Their metabolism gets increasingly complicated and takes considerably longer when the degree of branching grows.

Their usage as a feedstock for the synthesis of single cell protein (SCP) has received a lot of attention, and numerous large-scale alkane-based SCP factories have been established. The largest facility has been shut down due to issues about harmful ingredients and expenses.

Alkenes

Chemical compounds composed solely of carbon and hydrogen, with at least one unsaturated (double) >C=C bond. As a result, they are chemically reactive. Ethene (ethylene), which is used to make polythene, is a popular alkene utilised in commercial procedures.

Even though most polythene is made using fossil fuels, a handful of facilities have been established that use fermentation ethanol as a raw source.

Allele

Alternative form of genes that are found at the same locus on homologous chromosomes, are divided during meiosis, and are recombined after gamete fusion. When two are present, one (the dominant) is asserted above the other (the recessive).

Allergen

An antigen that causes the immune system to overreact.

Allopatric Speciation

The emergence of new species among populations that are geographically separated.

Allopolyploid

A reproductive organism with more than two chromosomal sets due to interbreeding and chromosome fusion between two distinct species of same genus.

Allosteric Regulation

The binding of a regulatory molecule and a protein at one location that alters the protein's function at another location.

Alpha (A) Helix

A coiled area of a protein's secondary structure that results from a particular arrangement of hydrogen bonding between atoms in the polypeptide backbone (not the side chains).

Alternation Of Generations

A life cycle in which a multicellular diploid form, the sporophyte, coexists with a multicellular haploid form, the gametophyte.

Alternative RNA Splicing

A kind of eukaryotic gene regulation at the RNA-processing level wherein various mRNA molecules are generated from the very same basic transcript relying on which RNA segments are considered as exons and spliced region as introns.

Altruism

Selflessness. Defined as an activity that diminishes one's fitness but boosting the fitness of another.

Alveolate

A protist having alveoli (membrane-bounded sacs) immediately under the plasma membrane.

Alveolus

In a mammalian lung, one of the dead-end air sacs wherein gaseous exchange takes place.

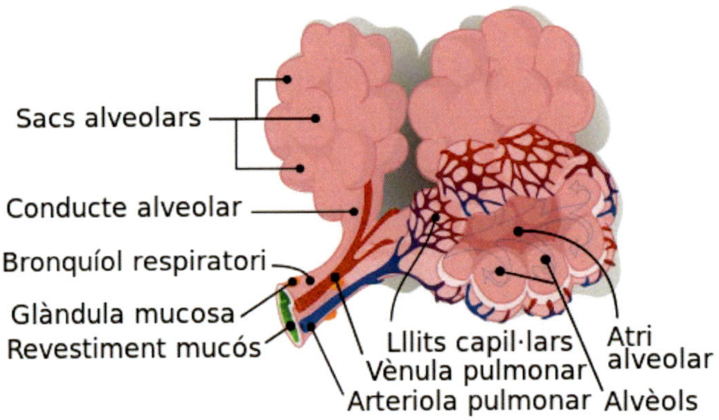

Figure 9. Alveolus Diagram.

Source : Image By Wikimedia Commons

Alzheimer's Disease

A kind of dementia (mental degeneration) caused by ageing that is marked by confusion and loss of memory.

Amacrine Cell

A retinal neuron that aids in the integration of information before it can be transmitted to the brain.

Amino Acid

An organic molecule possessing both a carboxyl and an amino group. Amino acids serve as the monomers of polypeptides.

Aminoacyl-tRNA Synthetase

An enzyme that connects each amino acid to its corresponding tRNA.

Ammonia

A tiny, poisonous molecule (NH3) formed as a metabolic waste product of protein and nucleic acid metabolism or as a nitrogen fixation product.

Ammonite

A member of a group of shelled cephalopods that were significant marine predators for hundreds of millions of years before becoming extinct at the end of the Cretaceous era (65.5 million years ago).

Amniocentesis

A prenatal diagnostic procedure wherein amniotic fluid is acquired by aspiration from a needle put into the uterus. The fluid and the foetal cells in it are tested to see if the foetus has any genetic or congenital abnormalities.

Amniote

A member of a tetrapod clade termed after a significant derived trait, the amniotic egg, which comprises specialised membranes that protect the embryo, such as the fluid-filled amnion. Mammals, birds, and other reptiles are all amniotes.

Amniotic Egg

An egg with specific membranes that help with protection, nutrition, and gaseous exchange. The amniotic egg was a significant evolutionary advance because it enabled embryos to grow on land in a fluid-filled sac, eliminating tetrapod's reliance on water for reproduction.

Amoeba

A kind of protist distinguished by the presence of pseudopodia.

Amoebocyte

An amoeba-like cell present in all animals that travels using pseudopodia. It may digest and distribute food, get rid of wastes, form skeletal fibres, fight infections, or shift into other cell types, based on the species.

Amoebozoan

A protist with lobe- or tube-shaped pseudopodia that belongs to a clade that comprises multiple species.

Amphibian

A member of the tetrapod class amphibia, which includes salamanders, frogs, and caecilians.

Amphipathic

The presence of both a hydrophilic and a hydrophobic zone.

Amplification

The increase in stimulus energy during transduction.

Amygdala

A structure in the vertebrate brain's temporal lobe that plays a key function in emotion processing.

Amylase

An enzyme that breaks down starch (a plant-derived glucose polymer) and glycogen (an animal-derived glucose polymer) into smaller polysaccharides and the disaccharide maltose.

Anabolic Pathway

A metabolic process that uses energy to make a complex compound from simpler ones.

Anaerobic Respiration

A catabolic route in which inorganic substances besides oxygen receive electrons at the "downhill" end of electron transport chains.

Analogous

The presence of characteristics that are similar due to convergent evolution rather than homology.

Analogy

The commonality between two species resulting through convergent evolution rather than descent from a common ancestor who shared a property.

Anaphase

The fourth stage of mitosis, during which the chromatids of each chromosome have divided and the daughter chromosomes have travelled to the cell's poles.

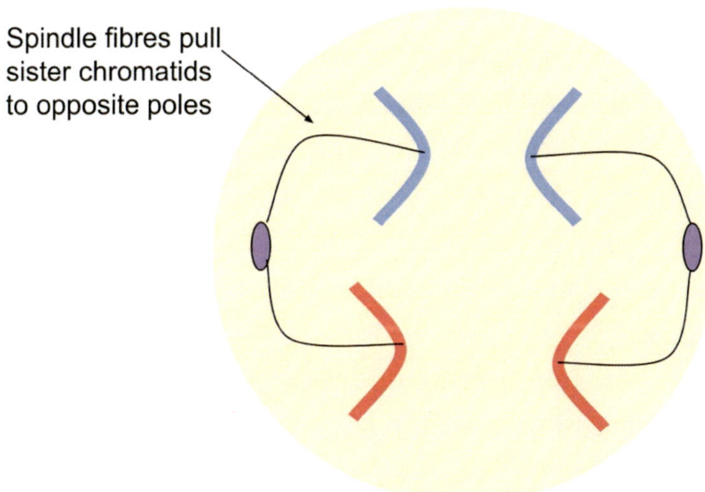

Figure 10. Anaphase During Mitosis.

Source : Image By Wikimedia Commons

Anchorage Dependence

The condition that a cell should be linked to a substratum before cell division can begin.

Androgen

A steroid hormone that promotes the growth and preservation of the male reproductive system and secondary sex characteristics, – for example testosterone.

Aneuploidy

A chromosomal abnormality wherein one or more chromosomes are duplicated or are missing altogether.

Angiosperm

A growing plant that produces seeds within a protected chamber known as an ovary.

Angiotensin II

A peptide hormone that causes precapillary arterioles to contract and enhances NaCl and water reabsorption by the proximal tubules of the kidney, thereby raising blood pressure and volume.

Anhydrobiosis

A dormant state characterised by the loss of nearly all bodily fluids.

Animal Pole

The point in the hemisphere where the least yolk is accumulated at the end of an egg. The opposite of vegetal pole.

Anonymous FTP

A file transfer protocol (FTP) that allows the retrieval of files from public sites.

Annotation

A collection of comments, notations, references, and citations, either in free format or utilizing a controlled vocabulary, that together describe all the experimental and inferred information about a gene or protein. Annotations can also be applied to the description of other biological systems. Batch, automated annotation of bulk biological sequence is one of the key uses of bioinformatics tools.

Anterior

Relating to a bilaterally symmetrical animal's front, or head.

Anterior Pituitary

A region of the pituitary gland that arises from nonneural tissue and is composed of endocrine cells that manufacture and produce a variety of tropic and non-tropic hormones.

Anther

The terminal pollen sac of a stamen of an angiosperm, in which pollen grains with sperm-producing male gametophytes develop.

Antheridium

The male gametangium, a wet chamber wherein the gametes grow in plants.

Anthropoid

A member of the primate family that includes monkeys and apes (gibbons, orangutans, gorillas, chimpanzees, bonobos, and humans).

Antibody

A protein that is released by plasma cells (differentiated b cells) and attaches to a specific antigen; often known as immunoglobulin. Antibodies always have the same y-shaped structure and are made up of two identical heavy chains and two identical light chains when monomerized.

Antibody array

An antibody microarray is a specific form of protein microarrays, a collection of capture antibodies are spotted and fixed on a solid surface such as glass, plastic or silicon chip, for the purpose of detecting antigens.

Anticodon

A nucleotide triplet on one end of a tRNA molecule that mates with a corresponding codon on a mRNA molecule.

Antidiuretic Hormone (ADH)

A peptide hormone that stimulates water retention by the kidneys (often referred to as vasopressin). Adhesion molecule (ADH) is formed in the hypothalamus and discharged from the posterior pituitary. It has a function in the brain.

Antigen

A chemical that triggers an immunological response by attaching to b cell, antibody, or t cell receptors.

Antigen Presentation

The process through which a MHC molecule attaches to a piece of an intracellular protein antigen and transports it to the cell surface, where it may be seen and identified by a t cell.

Antigen Receptor

A term used to denote a surface protein found on b and t cells that attaches to antigens and triggers adaptive immune responses. Antigen receptors on b cells are known as b cell receptors, whereas those on t cells are known as t cell receptors.

Antigen-Presenting Cell

A cell that produces peptide fragments that are bound by class ii MHC molecules and then presented on the cell surface to t cells after eating pathogens or internalising pathogen proteins. The principal antigen-presenting cells are macrophages, dendritic cells, and b cells.

Antiparallel

In a DNA double helix, the arrangement of the sugar-phosphate backbones. They run in opposite 5 s 3 directions.

Antisense technology

Molecular method that uses a nucleic acid sequence complementary to an mRNA so that the two bind and the mRNA is effectively neutralized.

Aphotic Zone

The area under the photic zone of an ocean or lake where enough light does not reach and therefore photosynthesis cannot take place.

Apical Bud

A bud at the apex of a plant stem. Another term for this is terminal bud.

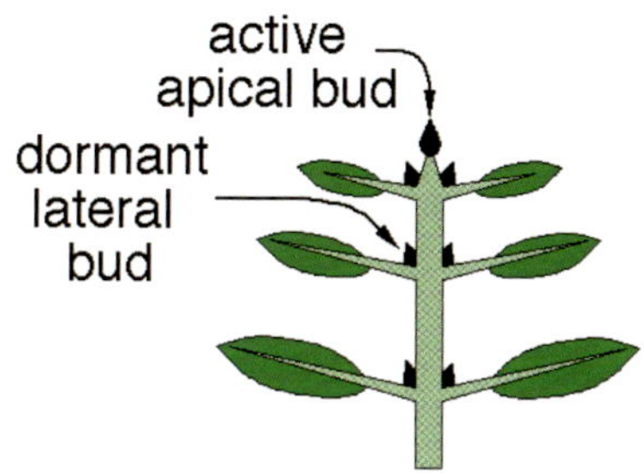

Figure 11. Apical Cycle.

Source : Image By Wikimedia Commons

Apical Dominance

The propensity for growth to be centralised at the apex of a plant shoot, since the apical bud limits axillary bud growth in part.

Apical Ectodermal Ridge (AER)

Embryonic plant tissue found at the roots' tips and in the buds of shoots. An apical meristem's dividing cells allow the plant to increase in length.

Apical Meristem

Embryonic plant tissue in the tips of roots and buds of shoots. The dividing cells of an apical meristem enable the plant to grow in length.

Apicomplexan

A parasitic protist belonging to a clade that comprises numerous different species. Human sickness is caused by some apicomplexans.

Apomixis

The capacity of some plant species for asexual reproduction using seeds and without the use of a male gamete.

Apoplast

Everything that is outside of a plant cell's plasma membrane, such as cell walls, intercellular gaps, and the space inside dead structures like xylem tubes and tracheid's.

Apoptosis

A form of planned cell death caused by the activation of enzymes that degrade several chemical components in the cell.

Aposematic Coloration

A vibrant warning coloration in many animals with excellent physical or chemical defences.

Appendix

A tiny, finger-like expansion of the vertebrate cecum that carries a mass of white blood cells that help the body fight infection.

Aquaporin

A channel protein in a plant, animal, or microorganism cell's plasma membrane that allows osmosis, or the flow of free water across the membrane.

Aqueous Solution

A solution wherein water functions as the solvent.

Arachnid

A member of the chelicerates, a prominent arthropod group. Spiders, scorpions, ticks, and mites are examples of arachnids.

Arbuscular Mycorrhiza

An association between a fungus and a plant root system wherein the fungus invades the plasma membranes of the host (plant) cells.

Arbuscular Mycorrhizal Fungus

A symbiotic fungus with hyphae that penetrate through the cell wall of plant roots and then into the root cell (enclosed in tubes formed by invagination of the root cell plasma membrane).

Archaea

One out of two prokaryotic domains (the second one is bacteria)

Archaeplastida

One of five eukaryote supergroups postulated in a recent eukaryote evolutionary history hypothesis. This monophyletic group contains red algae, green algae, and terrestrial plants, and evolved from an early protist ancestor that consumed a cyanobacterium.

Archegonium

The female gametangium, which is a wet chamber wherein gametes grow in plants.

Archenteron

The endoderm lined chamber that develops into an animal's digestive system during gastrulation.

Archosaur

A member of the reptile group containing crocodiles, alligators, and dinosaurs, as well as birds.

Array

An orderly pattern of objects. In genomic studies, there are microarrays and macroarrays. Microarrays are small spots of DNA or protein and the identity of the spotted material is known. Macroarrays are bacterial, yeast or similar colonies on plates used to determine functional consequences of genomic manipulations.

Arteriole

A blood vessel that transports blood from an artery to a capillary bed.

Artery

A blood artery that transports blood from the heart to organs all over the body.

Arthropod

A segmented ecdysozoan having jointed legs and a rigid exoskeleton. Insects, spiders, millipedes, and crabs are all common examples.

Artificial Selection

Selective breeding of domestic plants and animals in order to promote the appearance of favourable characteristics.

ASCII

Stands for American Standard Code for Information Interchange, sets a unique numeric definition for each of the most commonly used characters in Western language, including numerals, letters and some symbols. An ASCII set is these characters and their numbers. In the context of a file, an ASCII file is one that contains only text characters: numbers, letters and standard punctuation.

Ascomycete

Sac fungus, which is a member of the fungal phylum Ascomycota. The spores form in a saclike shape, which yields the name.

Ascus

A saclike spore capsule found at the apex of a sac fungus's dikaryotic hypha.

Asexual Reproduction

The production of offspring without the fusing of gametes from a single parent i.e. the offspring is produced either through budding, division of a single cell, or division of the entire organism into two or more parts. The offspring are almost always genetically similar to their parents.

Assembly

A compilation of overlapping sequences from one or more related genes that have been clustered together based on their degree of sequence identity or similarity. Sequence assembly may be used to piece together " shotgun " sequencing fragments based upon overlapping restriction enzyme digests, or may be used to identify and index novel genes from single - pass cDNA sequencing efforts.

Assisted Migration

The relocation of a species to a more suitable environment outside of its natural range in order to safeguard it from manmade hazards.

Assisted Reproductive Technology

A fertilisation treatment in which eggs (secondary oocytes) are surgically removed from a woman's ovaries following hormonal stimulation, fertilised, and returned to the woman's body.

Associative Learning

The developed ability to link an environmental element (like a colour) with some other environmental element (such as danger).

Aster

In a mitotic animal cell, a radial arrangement of small microtubules that spreads from each centrosome towards the plasma membrane.

Astrocyte

A glial cell that serves a variety of activities, such as giving structural support to neurones, controlling the interstitial environment, aiding synaptic transmission, and helping in the regulation of the brain's blood supply.

Atherosclerosis

A cardiovascular condition in which fatty deposits known as plaques form in the inner walls of the arteries, and thereby blocking and hardening them.

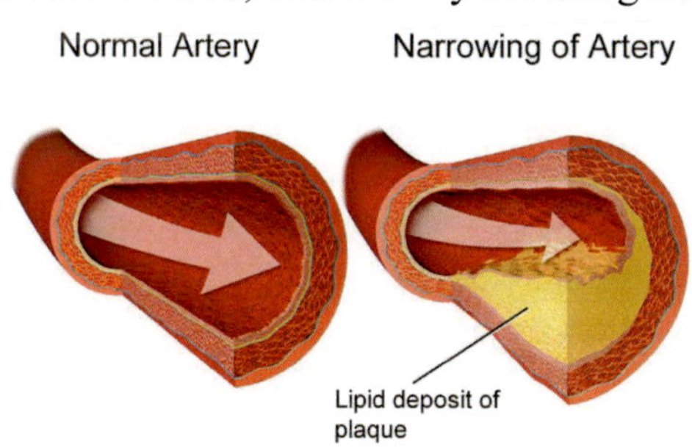

Figure 12. Atherosclerosis.

Source : Image By Wikimedia Commons

Atom

The smallest unit of a substance that preserves an element's characteristics.

Atomic Mass

The total mass of an atom. It is equal to the mass of 1 mole of the atom in grams.

Atomic Number

The amount of protons in an atom's nucleus, which is different for each element and is denoted by a subscript to the left of the element symbol.

ATP (Adenosine Triphosphate)

A nucleoside triphosphate that contains adenine and yields free energy when its phosphate bonds are hydrolysed. In cells, this energy is utilised to drive endergonic processes.

ATP Synthase

A combination of multiple membrane proteins that works in chemiosmosis with neighbouring electron transport chains to produce ATP by using the energy of a hydrogen ion (proton) concentration gradient. Eukaryotic cells have ATP synthases in their inner mitochondrial membranes, while prokaryotes have ATP synthases in their plasma membranes.

Atrial Natriuretic Peptide (ANP)

A peptide hormone released by heart atria cells as a response to increased blood pressure. The impact of ANP on the kidneys changes ion and water transport while also lowering blood pressure.

Atrioventricular (AV) Node

An area of specialised cardiac muscle tissue located between the left and right atria in which electrical impulses are paused for approximately 0.1 seconds prior to their dissemination to both ventricles and prompting them to contract.

Atrioventricular (AV) Valve

A cardiac valve that prevents blood backflow as the ventricle contracts and is placed between each atrium and ventricle.

Atrium

The chamber of the vertebrate heart that collects blood from veins and transports it to the ventricle.

Autocrine

A secreted substance that has an effect on the cell that released it.

Autoimmune Disease

An immunological condition wherein the immune system targets one's own body.

Autonomic Nervous System

The efferent section of the vertebrate peripheral nervous system that governs the internal environment. It is comprised of the sympathetic, parasympathetic, and enteric divisions.

Autopolyploid

An individual with more than two sets of chromosomes originated from a single species.

Autosome

A chromosome sets present in an organism that does not play a fundamental role in determining sex.

Autotroph

An organism that acquires organic food molecules without consuming other creatures or compounds generated from them. To produce organic molecules out of inorganic ones, autotrophs require energy from the sun or the oxidation of inorganic substances.

Auxin

A plant hormone that largely pertains to indoleacetic acid (IAA), which is a natural plant hormone that affects cell elongation, root development, secondary growth, and fruit growth, among other things.

Average Heterozygosity

The average percentage of loci in a population that are heterozygous in members of the population.

Avirulent

A term for a pathogen that can cause minor damage to the host but not death.

Axillary Bud

A structure with the ability to produce a lateral shoot or branch. The bud emerges at the intersection of a leaf and a stem.

Axon

A neuron's lengthy extension, or process, that transports nerve impulses out from the cell body to target cells.

B

B Cells

The lymphocytes that finish their development in the bone marrow and go on to become humoral immune effector cells.

Bacillus

A rod-shaped genus of aerobic spore-forming bacteria. Bacillus species are soil organisms that are chemoorganotrophic, that is they can utilize a wide range of organic compounds. Some of the species are pathogens. Most just require a few amino acids or vitamins as growth factors, and have modest dietary requirements.

The spore shape divides the species into three subgroups. In the first, the spore's diameter is never bigger than the cell's. The spore in the second group is oval and thick-walled, whereas the spore in the third group is spherical.

Bacillus thuringiensis

One of a variety of bacteria that generate protein crystals in connection with their dormant spores. These crystals are extremely poisonous to Lepidoptera (butterfly) larvae and related forms. The crystals are made up of a protein that is insoluble in acid but dissolves in alkaline environment, such as those found in the larval stomach.

The dissolved protein damages the gut cell wall, causing alkaline contents to diffuse into the insect's bloodstream, producing paralysis. Any spores taken at the same moment may germinate, infect the larva's tissue, and kill it.

Protein crystals and spores from a variety of *B. thuringiensis* strains are being used as pest control agents, and they are efficient against a variety of caterpillars and mosquito larvae.

Bacitracin

Bacillus licheniformis produces a peptide having antibiotic properties.

Back Cross

A method of crossing progeny (F1) with one of the parents. It's a test that determines if a person is homozygous or heterozygous for a certain gene.

The back cross may be used to separate the phenotypically identical AA and Aa, because the cross between the homozygous dominant and the homozygous recessive parent will result in offspring that are all Aa and display the dominant feature. In contrast, a cross between a heterozygous F1 and a double recessive parent will result in 50% of the children having the recessive trait.

Back Mutation

A mutation that restores the function of a mutant gene to that of its wild-type counterpart.

Bacteria

A Protista division made up of single-celled prokaryotes. The Monera is made up of bacteria and cyanobacteria (blue-green algae). These bacterias are one of the smallest creatures with a cellular structure, with an average diameter of around 1 J..tm and lengths ranging from 0.1 to 10 J..tm. True bacteria (eubacteria), actinomycetes, chlamydobacteria, beggiatoa, myxobacteria, spirochaetes, mycoplasms, and rickettsias are some of the subgroups.

These organisms are classed as coccus, bacillus, spirillum, or vibrio based on their form. The cells are surrounded by a stiff cell wall (which may be encircled by a capsule or slime layer) and a plasma membrane that encloses the cytoplasm, which lacks membrane-bound organelles seen in eukaryotic cells.

Ribosomes are present in the nuclear apparatus, which is made up of a circular DNA molecule. The photosynthetic apparatus is connected with particular membranes or membrane-bound vesicles, rather than chloroplasts, in photosynthetic animals. Flagella can be found on one or more motile species.

The majority are saprophytes, but some are pathogens of plants, people, other animals, and microbes. Bacteria have a diverse set of nutritional needs and energy metabolism. Some parasite species will only develop on a medium of living animal tissue, while others require sophisticated mediums including a diverse range of amino acids, vitamins, and cofactors.

Bacteria can be aerobic, anaerobic, or facultative anaerobes, depending on their environment. Bacteria play a significant role in soil organic matter decomposition, nitrogen fixation, and rumen metabolism. Bacteria are utilized in waste disposal, food processing, certain fermentations, anaerobic digestion, and antibiotic manufacturing, as well as in the creation of industrial enzymes and bulk and fine compounds.

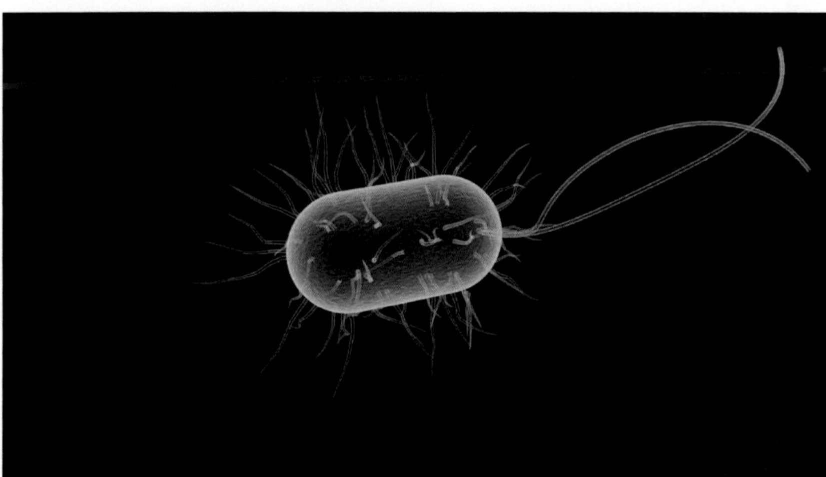

Figure 13. Image Showing a Bacterium.

Source : Image By Pixabay

Bacterial Artificial Chromosome (BAC)

A plasmid that may hold inserts of 100,000 to 300,000 base pairs (100–300 kb) that operates as a bacterial chromosome.

Bacterial Chromosome

A single DNA molecule that can be up to 1 mm long and has a molecular weight of more than 107 daltons, containing 4000-5000 kilobases.

Bacterial Community

A bacterial culture that contains populations from a variety of distinct species.

Bacterial Population

A collection of bacteria from the same species that exchange genetic material but have little interaction with other bacteria from the other species.

Bacteriochlorophyll

In photosynthetic bacteria, this chlorophyll replaces chlorophyll A.

Bacteriocide

A poisonous substance that can kill bacteria. Common disinfectants like phenol and bleach (hypochlorite), as well as mercuric chloride and organic mercury compounds, are examples of bacteriocides.

Bacteriocin

Any of the proteinaceous toxins produced by a variety of bacterial strains. These chemicals are frequently called by the organism that produces them, such as colicins, which are generated by *E. coli*, and staphyloccins, which are produced by *Staphylococcus* species.

Bacteriocins are potent against strains that are closely similar, but the cells that produce them are immune. Instead of being encoded in the main bacterial genome, these proteins are encoded by plasmid genes.

Bacteriolysis

Disolution of bacterial death.

Bacteriophage

A virus that infects bacteria by inserting its nucleic acid into the host's DNA. Only three genes are coded for on single-stranded RNA in the smallest phage. Larger DNA-containing bacteriophages, such as T4, which uses E. coli as a host, have a molecular weight of roughly 2.2 x 108 daltons and carry around 60 genes on a double-stranded DNA genome wound inside the phage's head.

Infection occurs when the phage attaches to the bacterial surface and the sheath contracts, forcing the tail core through the wall. Following that, the nucleic acid is introduced into the bacteria via the hollow tail. In contrast to viruses that infect eukaryotic cells, bacteriophages seldom enter their hosts as complete particles.

Infection can lead the host cell's metabolism to turn fully to the production of new bacteriophage, resulting in the host's lysis and the release of new virus particles. Alternatively, viral DNA might be incorporated into the host chromosome, remain latent, and proliferate as a prophage with the host chromosome.

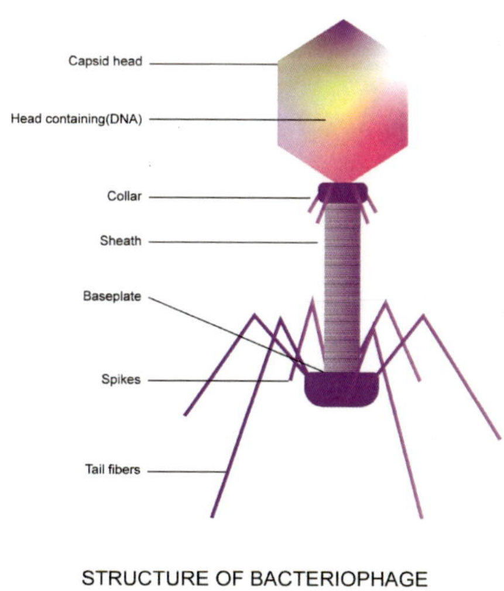

Figure 14. Structure of a Bacteriophage.

Source : Image By Wikimedia Commons

Bacteroid

Gram negative, anaerobic bacteria found especially in root nodules of leguminous plants. Takes part in N_2 fixation.

Baculovirus

Only arthropods are infected by this category of viruses. They're divided into groups based on whether or not the virions get embedded (occluded) in inclusion bodies, as well as the shape and size of these structures. Lepidoptera, Diptera, Hymenoptera, and Coleoptera are infected by nuclear polyhedrosis viruses, which have polyhedral inclusion bodies.

Only Lepidoptera are infected by granulosis viruses that get occluded in capsule-shaped inclusion bodies, but those that do not form inclusion bodies (nonnoccluded) infect a wide spectrum of arthropods. These viruses are interesting as possible biological pest management agents because of their host selectivity, simplicity of detection, and the fact that they do not infect vertebrates.

Bagasse

The solid leftovers left after sugar cane juice has been squeezed. It is used as a fuel to provide energy for cane crushing and the distillation of alcohol after cane juice or molasses fermentation.

Bakers' Yeast

Saccharomyces cerevisiae strains produced under extremely aerobic conditions utilizing pure culture techniques using cane or beet molasses as substrate. The yeasts are collected by centrifugation, which yields a cream with about 20% solids that is pressed in a filter press or filtered in a rotary vacuum filter to yield compressed yeast.

Alternatively, you can dry the yeast. In contrast to the yeasts used in brewing, distilling, and winemaking, these yeast types are not proprietary in general.

Breads

Bread is the most common baked food that incorporates microorganisms in its manufacture. Most breads are created with wheat flour and leavened using baker's yeast. The yeast releases carbon dioxide that is formed during the metabolism of carbohydrates in the dough, producing an aerated sponge-like structure that is fixed during baking.

The yeast also adds to the dough's flavor and maturation, especially through the creation of alcohol, which affects the colloidal state of the flour proteins.

Balanced Lethals

A condition in which numerous recessive deadly genes are carried at separate loci, resulting in at least one fatal gene on each homologous chromosome. These genes, however, are linked to inversions, preventing recombination between homologous chromosomes.

Balancing Selection

Natural selection in which two or more phenotypic types are maintained in a population.

Banding

After staining or under phase contrast microscopy, chromosomes show a pattern of contrasting bands. These bands are especially evident in metaphase chromosomes, where they can be strengthened by using enzymes and/or heat in combination with basic dyes. This type of chromosomal banding can be used to identify particular chromosomes within a pair.

BankIt

A computer program developed by the NCBI for submitting your own sequences to genbank.

Bark

External tissues of the vascular cambium, consisting primarily of secondary phloem and periderm layers.

Barr Body

Staining of the interphase nucleus of the homogametic sex (typically female) of some animals reveals a component of heterochromatin. Because the number of Barr bodies is one fewer than the number of X chromosomes, it can be used to detect disorders defined by sex chromosomal complement deficiencies or multiple complements.

An apparent female with just one X chromosome will have no Barr body, but an aberrant female with three X chromosomes will have two.

Basal Body

Microtubule triplets are arranged in a "9 + 0" pattern in eukaryotic cells. The basal body, which is physically similar to a centriole, may organize the microtubule assembly of a cilium or flagellum.

Basal Metabolic Rate (BMR)

At a moderate temperature, the metabolic rate of a resting, fasting, and nonstressed endotherm.

Basal Taxon

A taxon in a group of organisms whose evolutionary lineage separated early in the group's history.

Base

A chemical that lowers the concentration of hydrogen ions in a solution.

Base Analogue

A substance that is comparable to or a derivative of a nucleic acid base. Mutagens include analogues such as 5-bromouracil and 2-aminopurine. These might be integrated into DNA in place of thymine and adenine, resulting in erroneous base pairings in subsequent cell generations.

Basement Membrane

A thin coating of mucopolysaccharides and fibrous material between epithelial cells and the underlying tissue that functions as a barrier.

Base Pairing

In nucleic acids, the contact between opposing purine and pyrimidine bases that happens through hydrogen bonding. Between adenine and thymine (A + T) and guanine and cytosine (G + C), stable base pairs are created.

All interactions between nucleic acids and the transfer of information from one molecular species to another, such as the synthesis of RNA on a DNA template or the interaction between complementary nucleotide sequences of tRNA and mRNA in protein synthesis, are based on the specificity of base pairing. It also serves as the foundation for protein synthesis recognition and control signals.

Base Pairs

The hydrogen bonding that happens between complementary bases, such as A connecting with T and G binding with C, connects nucleotide pairs.

Base Ratio

The number of adenine and thymine bases (A + T) in double-stranded DNA compared to the amount of guanine and cytosine bases (G + C). The quantity of adenine is equal to the amount of thymine, and the amount of guanine is equal to the amount of cytosine as a result of the particular base pairing.

The ratio (A + T) to (G + C) is constant for a specific kind of DNA or for the total DNA in a given species, albeit it varies greatly between species. In bacteria, for example, it can range from more than 2 to less than 0.4.

Base-Specific Ribonuclease

An enzyme that breaks down an RNA molecule at a particular base: RNase Tl, for example, cleaves G residues, RNase U2 cleaves A residues, RNase PhyM cleaves both A and U residues, and RNase from Bacillus cereus cleaves U and C residues. These ribonucleases are used to analyse RNA sequences.

Basidiocarp

The higher basidiomycetes' fruiting body (e.g., mushrooms, toadstools and similar bodies).

Basidiomycetes

The most developed fungus, distinguished by the fact that spores are carried externally on a basidium within huge fruiting bodies. Mushrooms, puffballs, bracket fungi, rusts, smuts, stinkhorns, and other similar forms belong to this category.

Basidium

In basidiomycetes, a club-shaped structure generated from the terminal cell of a hypha that generates haploid basidiospores. The basidium produces sterigmata, which are pointed outgrowths. There are four in most species, but just two in others. To form the basidiospore, the tip of each sterigma expands. Two nuclei fuse in the juvenile basidium, followed by meiosis, to produce four nuclei, one of which travels into each basidiospore.

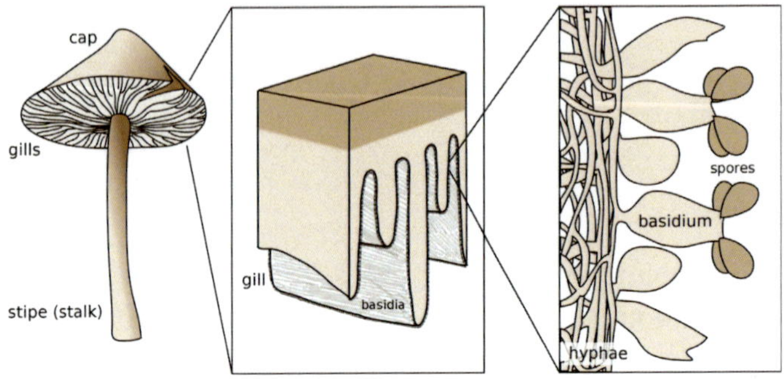

Figure 15. Basidium.

Source : Image By Wikimedia Commons

Basophil

A kind of white blood cell containing cytoplasmic granules that may be stained with basic dyes like methylene blue. Histamine is present in the cells, while heparin is present in the granules. They make up around half of the white blood cells in humans.

Batch Culture

A discontinuous system defined by a single loading or inoculation into new medium at the start of the process and a single harvesting step at the conclusion, generally after the substrate has been depleted, for cultivating microorganisms, waste treatment, or creating value-added products.

Following an initial lag period, organisms cultivated in batch systems often have a sigmoid growth curve, with exponential growth over a few generations. The cells may subsequently lyze, die, or stop growing as a result of the halt in growth.

Batch Recycle Culture

A discontinuous culture technique in which cells are collected at the conclusion of the growth phase and reintroduced into new media, resulting in the start of a new growth cycle. Because raw materials are not utilised to produce fresh cells for each cycle as in a traditional batch process, this approach is employed in the manufacturing of secondary products to maximise substrate usage efficiency.

Batesian Mimicry

A type of mimicry in which a non-poisonous species appears to be poisonous or otherwise dangerous to predators.

B Chromosomes

Accessory chromosomes are chromosomes that are not part of the usual complement. They can be found in a variety of plant and animal species, including maize and rye.

Beer

An alcoholic beverage made from malted grains fermented with yeast and additional hops. The mash is made by mixing ground malt with water and heating it to roughly 60°C, at which point the malt amylases liquefy the starch. This liquid is clarified and filtered to make a sweet wort, which is then cooked in copper pots with hops added.

This sterilizes the solution and inactivates the enzymes while also flocculating and removing various protein and phenolic contaminants. The sugar solution is concentrated to the required solids of roughly 12% during this procedure. The isomerization and hydrolysis of humulones, the bitter compounds contained in hops, are two essential reactions that happen at the same time.

The yeast is then added to the cooled wort. There are two main forms of beer, which are differentiated by the "yeast utilised." The first category includes lagers made with a bottom yeast (*Saccharomyces uvarum*, which flocculates and settles as the fermentation proceeds). The ales are made using the best yeast available (*S. cereviseae*, which rises to the surface during fermentation).

Behaviour

Individually, a response to a stimulus carried out by muscles or glands controlled by the neurological system; collectively, an animal's responses to external and internal stimuli.

Behavioural Ecology

The study of animal behavior's evolution and ecological underpinnings.

Benedict's Test

Using an alkaline copper sulphate solution and sodium citrate as a biochemical test for reducing carbohydrates. With increasing concentrations of sugar, reducing sugars produces a coloured precipitate that ranges from green to yellow to red.

Benign Tumor

A mass of aberrant cells with unique genetic and cellular alterations that prevent the cells from living in a new location and cause them to return to the tumor's original location.

Benthic Zone

An aquatic environment's bottom surface.

Benthos

The population of organisms that live in an aquatic biome's benthic zone.

Beta Oxidation

In this metabolic process, fatty acids are broken down into two-carbon pieces that enter the citric acid cycle as acetyl coa.

Beta (B) Pleated Sheet

The polypeptide chain folds back and forth in this type of secondary structure of proteins. Two sections of the chain run parallel to each other and are kept together by hydrogen bonds formed between the polypeptide backbone's atoms (not the side chains).

Bicoid

In *Drosophila melanogaster*, a maternal effect gene codes for a protein that specifies the anterior end.

Biennial

A plant that has a life cycle that spans two growth seasons (two years). The leaves are generally in the form of a prostrate rosette in the first season, and much of the nutrients is stored in an underground root that survives the winter. The stored material is utilized to form an upright, blooming stem in the second year, which subsequently produces seeds.

Bilateral Symmetry

Only one plane, generally going longitudinally through the midline, separates an item into two halves that are mirror reflections of one another.

Bilaterian

A member of a clade of animals with three germ layers and bilateral symmetry.

Bile

Cholesterol, bile pigments (biliverdin and bilirubin), bile salts (sodium salts of cholic acid coupled with taurine or glycine), and lecithin are all found in the secretions of the vertebrate liver. Through the bile duct, it is released into the small intestine.

Binding Site

An ion, compound, antibody, virus, cell, or organism recognises a certain arrangement of atoms or molecules and uses it to connect to another structure or creature.

Binary Fission

"Division in half" is an asexual reproductive process. Binary fission in prokaryotes does not entail mitosis, but mitosis is a part of the process in single-celled eukaryotes which go through binary fission.

Binomial

A binomen is a two-part, latinized method for naming a species that includes the genus and specific epithet.

Binomial Nomenclature

Linnaeus developed a method for naming plants and animals in the 18th century. A generic name and a species name are assigned to each species. After the species name, the discoverer's name is printed in abbreviated form. The International Codes of Nomenclature specify the specific technique and standards for naming new species.

The International Code of Nomenclature for Cultivated Plants, the International Code of Botanical Nomenclature, and the International Code of Zoological Nomenclature are all examples of these codes. When a new creature is discovered, it is given a Latin name, a description in the same language is written, and a holotype or type specimen is preserved.

In general, the organism's earliest or oldest name takes priority. Despite the fact that a particular animal or plant must have a name that is unique to the kingdom, a plant and an animal have been given the same name on multiple occasions. Other anomalies include the placement of phases of an organism's life cycle in separate genera or even phyla.

Bioassay

A method of assessing a substance's amount or concentration by observing how it affects a live organism under controlled settings. Bioassays are frequently used to determine compounds that exist in extremely low quantities, such as vitamins, hormones, and plant growth factors, because such procedures are many orders of magnitude more sensitive than standard chemical studies.

Biodiversity Hot Spot

A small area having a considerable number of indigenous species as well as endangered and vulnerable species.

Bioenergetics

1. An organism's total energy flow and transformation. 2. The study about how energy moves via living organisms

Biofilm

A surface-coating colony of one or more prokaryote species that collaborate metabolically.

Biofuel

A fuel made from dry organic matter or plant-produced combustible oils.

Biogeochemical Cycle

Any of the several chemical cycles in ecosystems that incorporate both biotic and abiotic components.

Biogeography

The study of species' geographic distribution in the past and now.

Bioinformatics

The processing and integration of biological data from massive data sets using computers, software, and mathematical models.

Biological Augmentation

A method of restorative ecology that involves the employment of organisms to replenish critical components in a deteriorated environment.

Biological Clock

Internal clock that regulates an organism's biological cycles. The biological clock can keep track of time with or without external cues, although it frequently relies on signals from the outside world to stay on track.

Biological Magnification

A process in which retained chemicals become more concentrated at each higher trophic level in a food chain.

Biological Species Concept

A species is defined as a group of populations whose members have the ability to interbreed in nature and produce viable, fertile offspring, but not with members of other similar groups.

Biomanipulation

A technique to changing ecosystem traits that uses a top-down concept of community organization. Ecologists, for example, can avoid algae blooms and eutrophication in lakes by changing the density of higher-level consumers rather of employing chemical treatments.

Biomass

A group of organisms in a specific habitat's total mass of organic matter. Any of the world's primary ecosystem types, commonly categorized by the dominating flora for terrestrial biomes and the physical environment for aquatic biomes, and characterized by organism adaptations to that specific habitat.

Bioremediation

Detoxification and restoration of polluted and damaged habitats using organisms.

Biosphere

The total of all the planet's ecosystems; the entire region of the globe inhabited by life.

Biotechnology

The process of manipulating organisms or their components in order to create useful goods.

Bipolar Cell

A neuron in the retina that transfers information between photoreceptors and ganglion cells.

Bipolar Disorder

A depressed mental condition marked by mood fluctuations from high to low; sometimes known as manic-depressive disorder.

Birth Control Pill

A chemical contraception that prevents sperm from accessing the uterus by inhibiting ovulation, delaying follicular development, or altering a woman's cervical mucus.

Blade

A seaweed's leaf like structure that supplies the majority of the surface area for photosynthesis. A typical leaf's flattened section.

BLAST

The protein and nucleic acid sequence search engine developed at NCBI that allows you to search sequence databases. Blastn searches for nucleotide sequences, blastp searches for amino acid sequences; BLAST2 compares 2 sequences.

Blastocoel

The chamber in the core of a blastula that is filled with fluid.

Blastocyst

The blastula stage of mammalian embryonic development is made up of an inner cell mass, a cavity, and a trophoblast outer layer. The blastocyst develops one week after conception in humans.

Blastomere

An early embryonic cell that develops during the cleavage stage of a developing embryo.

Blastopore

The opening of the archenteron in a gastrula, which in deuterostomes develops into the anus and in protostomes into the mouth.

Blastula

A hollow ball of cells that signals the conclusion of the cleavage stage in early embryonic development in mammals.

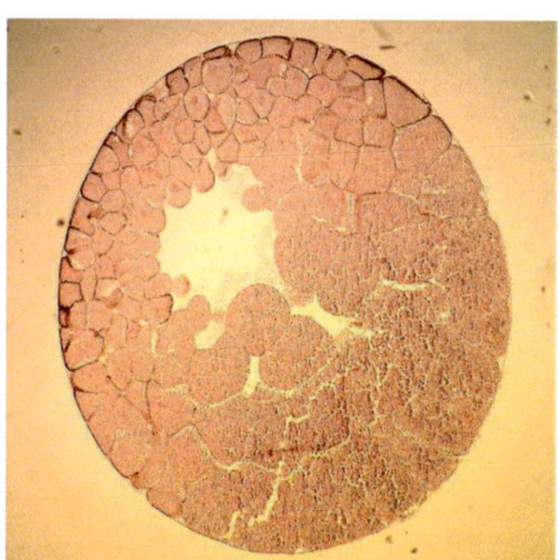

Figure 16. Prepared Slide of Ambystoma in the Blastula Stage.

Source : Image By Wikimedia Commons

Blood

A connective tissue in which red blood cells, white blood cells, and cell fragments called platelets are suspended in a fluid matrix called plasma.

Blue-Light Photoreceptor

Phototropism and hypocotyl elongation slowing are two examples of reactions triggered by this type of light receptor in plants.

Body Cavity

Between the digestive tract and the body wall, there is a fluid or air-filled gap.

Body Plan

A combination of morphological and developmental features that are integrated into a functioning whole—the live organism—in multicellular eukaryotes.

Bohr Shift

A dip in pH causes a decrease in hemoglobin's affinity for oxygen. It makes it easier for hemoglobin to release oxygen in the proximity of active tissues.

Bone

A connective tissue made up of live cells that are kept together by a hard matrix of collagen fibres embedded in calcium salts.

Book Lung

In spiders, a gas exchange organ made up of stacked plates housed in an interior chamber.

Bootstrap analysis

A statistical technique used to assess the reliability of the branching structure in phylogenetic trees.

Bottleneck Effect

Genetic drift that happens when the number of a population is diminished, such as by a natural calamity or human activity. The surviving population is typically not genetically representative of the initial population.

Bottom-Up Model

Mineral nutrients regulate community organization through influencing plant or phytoplankton populations, which in turn influence herbivore numbers, and subsequently predator numbers, according to this hypothesis.

Bowman's Capsule

The cup-shaped receptacle in the vertebrate kidney that receives filtrate from the blood and is the first, enlarged segment of the nephron.

Brachiopod

A marine lophophorate with a shell split into dorsal and ventral parts; sometimes known as a lamp shell.

Brain

Information is processed and integrated in this central nervous system organ.

Brainstem

A group of structures in the vertebrate brain, comprising the midbrain, pons, and medulla oblongata; roles in homeostasis, movement coordination, and information transmission to higher brain areas.

Branch Point

The separation of two or more species from a common ancestor as depicted on a phylogenetic tree. A branch point is typically depicted as a dichotomy, where a branch representing the ancestral lineage breaks into two branches, one for each of the two descendent lineages (at the branch point).

Brassinosteroid

In plants, a steroid hormone that causes cell elongation, delays leaf abscission, and promotes xylem differentiation, among other things.

Breathing

Inhale and exhale in a cycle to ventilate the lungs.

Bronchiole

The bronchial branch that delivers air to the alveoli.

Bronchus is one of two breathing tubes that branch from the trachea and enter the lungs.

Brown Alga

A photosynthetic multicellular protist with a distinctive brown or olive hue due to carotenoids in its plastids. The majority of brown algae are marine, with some having a plant-like body (thallus).

Bryophyte

A nonvascular plant that survives on land but lacks some of the terrestrial characteristics of vascular plants, such as moss, liverwort, or hornwort.

Budding

Asexual reproduction in which offshoots from the parent develop and pinch off to live independently or stay connected to create large colonies.

Buffer

A weak acid and its matching base are combined in a solution. When acids or bases are introduced to a solution, a buffer helps to keep the ph stable.

Bulk Feeder

An animal that consumes vast amounts of food

Bulk Flow

The movement of a fluid caused by a pressure differential between two points.

Bundle-Sheath Cell

A form of photosynthetic cell found in c4 plants that is distributed in densely packed sheaths around the veins of a leaf.

Buoyant Density

When a molecule is exposed to the force created in an analytical centrifuge, it achieves equilibrium density inside a density gradient established from a solution of a substance such as caesium chloride. The amount of cytosine plus guanine enhances the buoyant density of DNA.

Buttermilk

Streptococcus lactis, S. cremoris, and/or *Leuconostoc cremoris* cultures with pasteurised skimmed milk produce a fermented product.

C

C3 Plant

The bulk of photoautotrophs that produce oxygen. Carbon dioxide absorbed during photosynthesis is immediately fixed by the enzyme ribulose bisphosphate carboxylase and then metabolized through the photosynthetic carbon reduction cycle in C3 plants. The word c3 refers to the three-carbon compound 3-phosphoglyceric acid, which was discovered as the first stable result of carbon metabolism in radio labelling tests.

C3 plants include most aquatic plants, trees, shrubs, and temperate herbaceous plants. Oxygen inhibits photosynthesis in C3 plants. Low amounts of the enzyme phosphoenolpyruvate (PEP) carboxylase in green leaves distinguish them. They also exhibit high rates of photorespiration and carbon dioxide compensation points of roughly 40-100 ppm.

They demonstrate a -30 percent selectivity against the heavy carbon isotope (13C). Short-term photosynthetic yields range between 20 and 30 grams per square meter per day, with annual productivities ranging between 10 and 30 tons per hectare per year.

C4 Plant

Higher plants with a carbon absorption route in addition to the regular photosynthetic carbon reduction cycle. The designation C4 comes from the fact that the four-carbon organic acid malic acid and the four-carbon amino acid aspartic acid were the first stable compounds detected in the light using 14C labelling.

Although some temperate grasses and dicotyledons have c4 photosynthesis, the majority of C4 plants are tropical or subtropical grasses. A combination of morphological, anatomical, physiological, and biochemical characteristics, all of which must be present for it to be a real C4 plant, may be used to identify C4 plants.

Around the vascular bundles, photosynthetic tissue is organized in two concentric layers. The photosynthetic carbon reduction cycle enzymes, particularly ribulose bisphosphate carboxylase/oxygenase (rubisco), are limited to the inner bundle sheath layer.

Phospho-enol pyruvate (PEP) carboxylase and chloroplast enzymes involved in the production of PEP and the reduction or transamination of oxaloacetic acid to malate or aspartate are abundant in the outer, mesophyll layer. From the mesophyll layer to the bundle sheath layer, C4 acids are translocated.

The fixed carbon from the acids is then released and re-fixed in the bundle sheath through the usual carbon reduction cycle. Pyruvate or alanine, a three-carbon residue, is returned to the mesophyll layer and utilized to replenish PEP. As a result, the complete C4 cycle operates as a pump, boosting the concentration of carbon dioxide in the rubisco area.

The chloroplasts in most c4 plants are dimorphic, with the bundle sheath having bigger chloroplasts that preferentially collect starch. Bundle sheath chloroplasts in extreme C4 plants like sugar cane, maize, and sorghum exhibit significantly decreased grana and unusually low photosystem II activity. As a result, the rate of oxygen generation is slowed, and the oxygen/carbon dioxide ratio falls.

As a result, C4 plants are more resistant to oxygen inhibition than C3 plants. They have decreased photorespiration and carbon dioxide compensation values that are close to zero. They also show less selectivity towards the heavy carbon isotope (13C) than C3 or CAM plants, with levels of roughly - 15%.

In C4 plants photosynthesis occurs at the greatest rates, up to 80 grams per square meter per day, resulting in extremely high annual yields of up to 80 tons dry weight per hectare. The biochemistry of C4 plants can be categorized based on anatomical distinctions and minute features of their biochemistry.

The first is based on the type of the primary C4 acid produced, which is classified as malate or aspartate formers. The enzyme involved in decarboxylation of the C4 acid in the bundle sheath is classified at the second level. NADP-malic enzyme, NAD-malic enzyme, and PEP carboxykinase species can all be identified on this basis.

Calcitonin

A polypeptide hormone that inhibits parathyroid hormone and decreases plasma calcium levels. The C cells of the thyroid or parathyroid glands in mammals, as well as the ultimobranchial bodies of lesser vertebrates, release this hormone.

Calibration

The adjustment of an instrument to produce an absolute result in future determinations using legitimate standards or established methodologies. Standards with known chemical composition, molecular weight, or density are used to calibrate chromatographic, electrophoretic, and centrifugal procedures.

Callose

The sieve plate walls of sieve tubes include a 13-(1,3)-linked glucose polymer. It's deposited in vast quantities to seal off sieve plates during maturation, seasonal changes, or injury.

Callus

Undifferentiated (parenchymatous) tissue that is actively developing in higher plants in response to trauma and infection. A callus can also occur *in vitro* when plant tissue is artificially cultured.

Calmodulin

A eukaryotic regulatory protein that binds to a multitude of enzyme groups involved in cyclic nucleotide transport, calcium ion transport, contractile activities, neurotransmission, and certain phosphatases with total specificity.

This type of binding is directly dependent on the availability of calcium ions. This property allows calmodulin to be employed as an affinity product to or from which the above-mentioned proteins may be selectively bound or eluted with a high degree of specificity.

Calorie (Cal)

The heat energy necessary to raise the temperature of 1 g of water by 1°C, as well as the heat energy released when 1 g of water cools by 1°C. A kilocalorie (with a capital k) is a calorie (with a capital c) that is commonly used to represent the energy content of food.

A unit of energy equal to 4.1868 joules, or the amount of heat necessary to raise the temperature of one gram of water by one degree Celsius. The calorie, often known as the kilocalorie, is a unit of measurement for the energy content of foods. One kilocalorie equals 1000 calories.

Calvin Cycle

Fixation of atmospheric CO2 and reduction of the fixed carbon into carbohydrate is the second of two key phases in photosynthesis (after the light reactions).

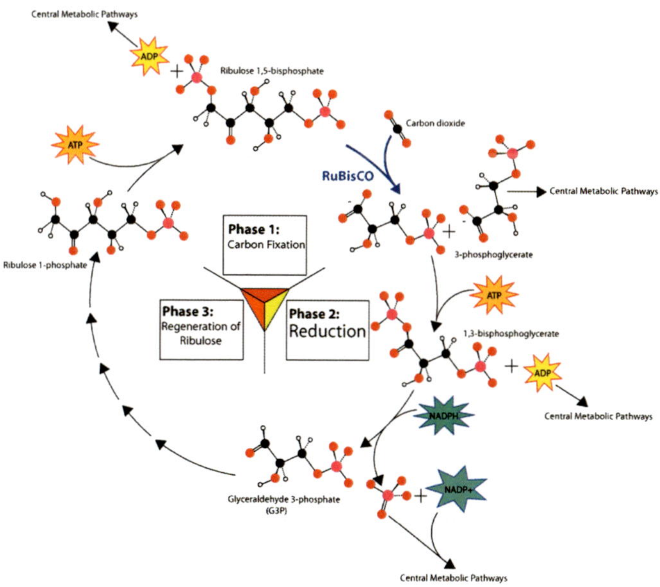

Figure 17. Calvin Cycle.

Source : Image By Wikimedia Commons

CAM Plant

A plant that employs crassulacean acid metabolism as a photosynthetic adaptation in dry environments. Carbon dioxide enters open stomata at night and is converted to organic acids, which release CO2 for the calvin cycle during the day when the stomata are closed.

Cambrian Explosion

Many modern-day animal phyla initially appeared in the fossil record within a very limited period in geologic history. Around 535–525 million years ago, a rush of evolutionary change occurred, resulting in the development of the first big, hard-bodied animals.

Canada Balsam

A gum that is dissolved in xylene and is commonly used to make permanent mounts of tissue slices and other specimens for microscopic study.

Cancer

Any of a variety of malignant tumors that can develop invasively and spread (metastasize) throughout the body via the lymphatic system and/or the bloodstream. On the basis of the kind of tissue in which they form, cancers are classified as carcinomas, lymphomas, sarcomas, and so on.

Canopy

In a terrestrial biome, the top layer of plants.

Capillary

A tiny blood artery that penetrates the tissues and is made up of a single layer of endothelial cells that permits blood and interstitial fluid to exchange.

Capping

The process by which nucleic acid terminal ends are covalently bonded to another type of molecule, rendering them chemically inert or even inactive. The capping of mRNAs' 5' ends *in vivo* makes them less vulnerable to phosphatases and nucleases. During *in vitro* polynucleotide synthesis, capping is also utilized to block the ends of partly produced intermediates.

Capsid

The protein shell that encases a viral genome. Its form might be rod-shaped, polyhedral, or more complicated. (1) A thick and well-defined covering of polysaccharide or protein that surrounds and is sticky, protecting the cell and allowing it to cling to substrates or other cells in many prokaryotes. (2) A bryophyte's sporangium (moss, liverwort, or hornwort).

Capsomere

A single polypeptide that constitutes a virus's capsid.

Capsule

A water-insoluble coating that builds up as loose, confluent layers outside the cell walls of many bacteria. Polysaccharides, often known as gums or slimes, are commonly used as coatings. Dextrans, levans, alginic acid, xanthan gum, and other carbohydrates can be found in carbohydrate capsules. A cellulose capsule is formed by other bacteria.

Some gram-positive bacteria, on the other hand, make polypeptide capsules made entirely of glutamic acid, while others also include amino sugars and uronic acids. Several of the gums are manufactured as industrial products by controlled fermentation and are used as gelling agents or thickeners.

Carbohydrate

A chemical composed of carbon, hydrogen, and oxygen, having the general formula Cx(H20)y. Depending on the degree of polymerization of the basic sugar units, carbohydrates are classified as monosaccharides, disaccharides, oligosaccharides, or polysaccharides.

They are the first products of photosynthesis and are used by animals, plants, and microbes as transport and storage materials. They also provide energy for development and movement in food, animal feed, and bacterial growth media. Monosaccharide phosphorylated derivatives are essential energy metabolism intermediates.

Carbon Balance

An investigation of the process efficiency of a fermentation or other bioreaction in which the distribution of carbon, given in the substrate, is assessed in relation to the distribution of cell biomass, carbon dioxide, and other metabolites on a theoretical or experimental basis.

Carbon Dioxide Concentration

The quantity of carbon dioxide in a gas or liquid phase represented in equivalent volume under normal circumstances, moles or weight per unit volume.

Carbon Fixation

An autotrophic organism's first assimilation of carbon from CO2 into an organic compound (a plant, another photosynthetic organism, or a chemoautotrophic prokaryote).

Carbon Source

The primary carbon source in a culture medium, providing the majority of the carbon structure for an organism's developing culture. Carbon dioxide is the most common carbon source for photosynthetic and autotrophic organisms. The carbon source for many heterotrophic organisms is generally glucose (produced from starch or cellulose) or sucrose.

Carbonyl Group

A chemical group that consists of a carbon atom double-bonded to an oxygen atom and is found in aldehydes and ketones.

Carboxylase

An enzyme that catalyzes a carbon dioxide incorporation process into another molecule. The key carboxylases in plants are ribulose bisphosphate carboxylase/oxygenase (rubisco), which catalyzes the primary absorption of carbon dioxide in photosynthesis, and phosphoenol pyruvate carboxylase, which is involved in auxiliary C4 metabolism in C4 photosynthesis and CAM plants.

Carboxyl Group

A chemical group found in organic acids that is made up of a single carbon atom that is double-bonded to an oxygen atom and also to a hydroxyl group.

Carcinogen

A cancer-causing agent that can be a chemical, ionising radiation, or a virus. The vast majority of carcinogens are also mutagens.

Cardiac Cycle

The phenomenon of the heart's alternating contractions and relaxations.

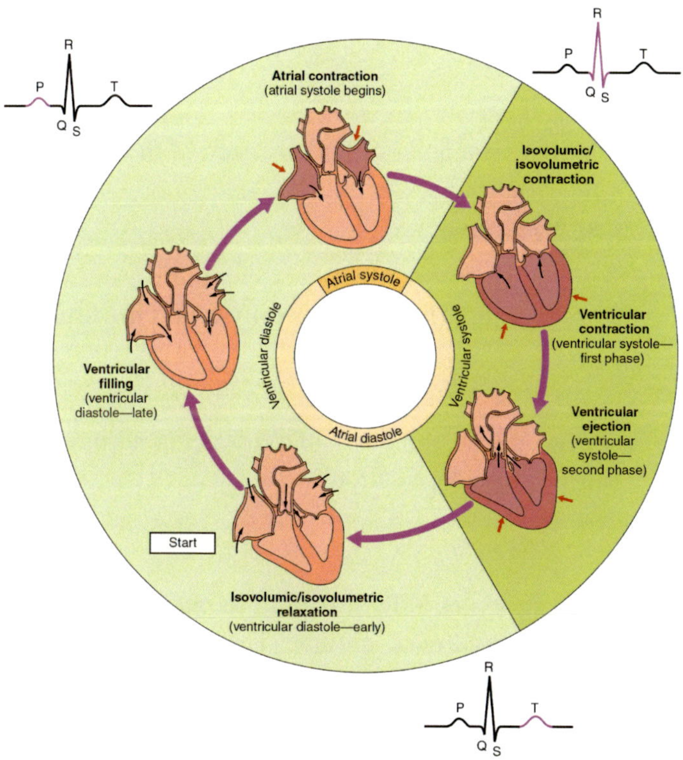

Figure 18. Phases Of Cardiac Cycle.

Source : Image By Wikimedia Commons

Cardiac Glycoside

A substance that has significant pharmacological effects on the heart muscle. Cardiac glycosides are made from plant tissue and contain a steroid aglycone that is either a C23 (cardenolide) or a C24 (bufadienolide) substance. They are formed by a joss of carbon atoms from the sterol's side chain.

These compounds' glycosidic portion is complex and contains unique sugars. Digitoxigenin from foxglove and hellebrigenin from the rhizome of the Christmas rose are among the aglycones.

Cardiac Muscle

A striated muscle that makes up the contractile wall of the heart. Intercalated discs connect its cells, relaying the electrical signals that underpin each heartbeat.

Cardiac Output

The amount of blood pumped per minute by each of the heart's ventricles.

Cardiovascular System

A heart plus a branching network of arteries, capillaries, and veins makes up a closed circulatory system. The system is only found in vertebrates.

Carnitine

A molecule involved in the transport of fatty acids across mitochondrial membranes during fatty acid oxidation.

Carotenes

Plant pigments that are structurally linked to tetraterpenes but lack oxygen functionalities such as hydroxyl groups. The presence of a long chain of conjugated double bonds gives these carotenoids their colour. As in B-carotene, cyclization can occur at either or both ends.

Carrots are the traditional source of "carotene," which is a composite of all carotenes, with B-carotene accounting for more than 90% of the total.

Carotenoids

Tetraterpene pigments are plant pigments with a tetraterpene structure. A variety of carotenoids are involved in photosynthesis as accessory pigments or protective agents, preventing chlorophyll photooxidation. The carotenes (hydrocarbon tetraterpenes) and xanthophylls (those containing oxygen functions, such as hydroxyl groups) are two types of carotenoids .

Carpel

The stigma, style, and ovary make up the ovule-producing reproductive organ of a flower.

Carrageenan

A gum that can be derived from Rhodophyceae algae. Carrageenans are polysaccharide mixes mostly composed of Dgalactose residues with 1,3)- and (1,4)- links. The galactose residues are mostly sulphated.

Carrier

A matrix that acts as a support for active biomass or cells in a bioreactor or cell culture system, allowing them to proliferate and become immobilised. Carriers are used to keep cell biomass in a fermenter or digester running continuously, or to encourage the growth and increase the production of animal cell lines that only grow when they come into touch with a suitable surface.

Carriers can range from big flat plastic plates to specially constructed plastic or ceramic rings, shattered rock, sand, or specially selected microporous glass spherules, depending on the system.

Carrying Capacity

The maximum population size that the available resources can support, denoted by the letter K.

Cartilage

Supportive tissue found in all vertebrates that is made up of a matrix that is deposited intercellularly with chondroblasts. There are several varieties of cartilage: hyalin cartilage, which is made up of chondroitin sulphate and collagen fibrils, and fibrocartilage, which is made up primarily of collagen and elastic fibres.

Cascade Fermentation

As the fermentation process progresses, the fermenting liquor or mash is fed through a series of fermenters. Overflow continuous fermentation is another name for the method. Up to ten fermenters can be connected in a series in industrial-scale systems. Sedimentation or centrifugation are used to remove the cells from the final fermenter.

These cells, which in the case of alcohol production are yeast, are reintroduced to the first fermenter, with any excess sold as animal feed or disposed of. If the amount of yeast recovered is insufficient, more growth is achieved by introducing air into the system or adding yeast from a starter culture tank.

To eliminate the physical removal and recycling of the cells, a variety of patented modifications have been devised, including techniques of holding or settling the yeast in intermediate tanks. This eliminates the need to acid wash the yeast, which is generally done in traditional procedures to limit the risk of infection during recycling. Fermentation times of 10 to 18 hours can be obtained using yeast.

Casein

A phosphoprotein, Milk casein is primarily made up of three types of proteins, one of which (x-casein) keeps casein micelles in solution and protects them from calcium ion flocculation.

Casparian Strip

A water-impermeable wax ring found in plant endodermal cells that prevents passive water and solute transfer into the stele via cell walls.

Catabolic

The breakdown of a complex biological molecule into less complex components is described by an enzymic process. Such processes can either produce ATP or metabolic intermediates that can be utilized in subsequent anabolic reactions.

Catabolism

The overall mechanisms involved in the breakdown of complex organic molecules into ATP. Catabolism includes processes such as glycolysis, glycogenolysis, proteolysis, and fatty acid oxidation.

Catabolite Gene Activator Protein (CAP)

A protein that binds with cyclic AMP close to the RNA polymerase binding site in the promoter region of an operon. The interaction of CAP with the promoter aids the development of the RNA polymerase-promoter complex, which acts as a positive regulator, allowing transcription to take place.

The addition of glucose lowers the amount of cyclic AMP and inactivates CAP, lowering transcription efficiency.

Catabolite Repression

A regulation mechanism in which glucose or glucose-derived metabolites block enzyme production pathways. Catabolite repression allows microorganisms to preferentially catabolize glucose when it is available.

This is due to the reduced activity of several bacterial operons in the presence of glucose (for example, the lac operon of *E. coli*). The active catabolism of glucose lowers the intracellular content of cyclic AMP, preventing an important positive regulatory signal for glucose-sensitive operon expression.

Catalase

The breakdown of hydrogen peroxide to oxygen and water is catalysed by this enzyme.

Catalysis

The insertion of a substance or enzyme (catalyst) that participates in, but is not permanently impacted by, a chemical reaction to promote it.

Catalyst

A substance that speeds up the rate of a chemical process while remaining unaltered at the conclusion. Enzymes function as biological catalysts.

Catastrophism

The idea that historical events happened unexpectedly and were caused by different processes than those in use now.

Catecholamine

Any of a class of neurotransmitters and hormones derived from the amino acid tyrosine, including the hormones epinephrine and norepinephrine.

Cathepsin

Cathepsins are proteases released from lysosomes responsible for cellular homeostasis.

Cation

In electrophoresis, a positively charged ion that is attracted to the cathode, or any positively charged ion, radical, or molecule.

Cation Exchange

A process in which positively charged minerals are made available to a plant when hydrogen ions in the soil displace mineral ions from the clay particles.

Cauliflower Mosaic Virus (CaMV)

A caulimovirus is a type of plant virus that affects Cruciferae and some Solanaceae. CaMV's genome is made up of an 8000-base-pair double-stranded circular DNA molecule.

The circular molecule has a number of single-stranded dicontinuities or gaps that occur at precise locations in respect to the restriction sites, which have been extensively mapped from several isolates. CaMV has been proposed as a potential vector for the introduction of foreign DNA into plant cells.

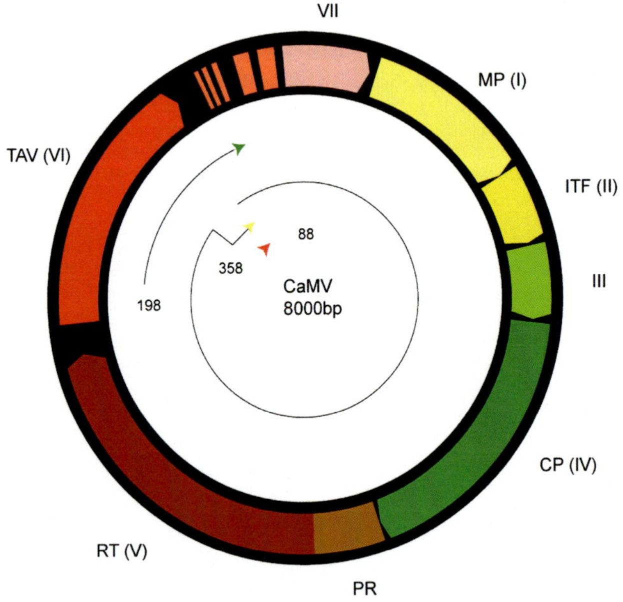

Figure 19. Cauliflower Mosaic Virus (CaMV).

Source : Image By Wikimedia Commons

Caulimo - viruses

Plant viruses with double-stranded DNA. The host range of these viruses is limited to a few closely related plants. They are primarily spread by aphids. Infection causes refractile spherical inclusions, which are made up of numerous viruses encased in a protein matrix.

The virus-coded protein can make up as much as 5% of the protein in infected cells. Cauliflower mosaic virus is the best-studied member of this group (CaMV).

cDNA

Complementary DNA is a single-stranded DNA that is created by the enzyme reverse transcriptase from an mRNA template.

cDNAs can be procured in a highly radioactive form and used as probes in hybridization assays to estimate the number and location of genes on a chromosome, to construct physical maps of different genomes, to study the organization of eukaryotic genes, and to locate coding sequences in a portion of DNA using radioactive nucleotides (generally 32P).

cDNA Library

A gene library that contains clones with complementary DNA (cDNA) insertion. Only the genes that were transcribed in the cells whose mRNA was recovered to create the cDNA are included in the library.

Cecum

The blind pouch is one of the big intestine's branches.

Cell

All living organisms' basic structural and functional unit. A plasma membrane surrounds the cytoplasm in all living cells. A hard or semi-rigid cell wall surrounds the majority of bacterial and plant cells.

By dividing or fusing previous cells, all cells are created. The cells contain both genetic material (DNA) that governs the inheritance of diverse characteristics and the machinery (ribosomes) that converts this information into proteins.

The organization of the genetic material in cells can be used to distinguish two types of cells: Prokaryotes have no nuclei and their genetic material is free in the cytoplasm; eukaryotes have a nucleus that separates the genetic material from the cytoplasm. The many enzymes required to catalyze all of the cell's anabolic and catabolic processes are found in the cytoplasm.

In prokaryotes, enzymes that conduct opposite roles are rarely separated, whereas in eukaryotes, enzymes or enzyme systems that catalyze opposing processes are housed in separate membrane-bound organelles.

These include organelles with a double membrane (chloroplasts, mitochondria, plastids, the nucleus, and so on) and those with a single membrane (lysosomes, peroxisomes, microbodies, spherosomes, Golgi bodies, and so on). Cells in multicellular animals can be specialized for a certain function and organized into tissues. Organs are made up of tissues that have similar functions.

Cell Body

The nucleus and most other organelles are housed in this region of the neuron.

Cell Breakage

Any mechanical process for breaking cells and releasing their contents. Pressure or mechanical shock, sonication, shaking, blending, ball milling or hammer milling, and other techniques are utilised.

Cell Counter

To enumerate the number of cells in a sample, an automated device is designed. To draw a known volume of a suitably diluted cell culture, the simpler type of cell counters consists of mechanism through a detection device. An electric potential is maintained while the detector comprises a plate with a small hole.

Using an electronic counter, a hole is amplified and logged, when any cell or particle passes through the hole is generated. Cell counters which are more sophisticated include flow cytometers.

Cell Counting

To enumerate the number of cells in a culture sample, a method is used. It can be done either manually or using an automated cell counter. Manual methods include direct or indirect. A direct count is done by diluting a suitable solution and counting the cells using a microscope and a hemocytometer, to get a statistical result.

Indirect methods require suitable dilution of the cells and plating out on nutrient agar plates, with the sufficient counting of the number of colonies that have shown on the plate. Generally, in the original culture one discrete colony will grow from each cell.

Cell Cycle

When a cell that has just undergone mitosis, here the growth cycle of an individual cell starts, thus has only one chromatid and half of the normal DNA content, the first phase is called as Gl.

Further comes the S phase in which each chromatid is replicated and this is followed by G2 phase where duplication process of other materials is completed. Then comes the mitosis process that occurs in the M phase.

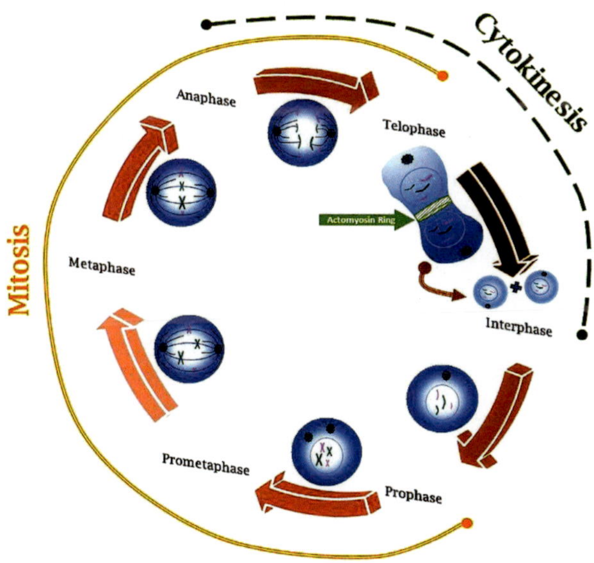

Figure 20. Cell Cycle Showing the Formation of the Actomyosin Ring During Telophase and Early Cytokinesis.

Source : Image By Wikimedia Commons

Cell Cycle Control System

The cell that functions both triggers and coordinates key events in the cell cycle is a cyclically operating set of molecules in the eukaryotic cell.

Cell Disruption

To liberate the content of cells a procedure is used. These procedures are mechanical and leads to cell breakage or based on cell lysis caused by addition of solvent that influence the cell membrane, antibiotics or anti metabolites which affects cell wall growth.

Cell Fractionation

The disruption of a cell and separation of its parts by centrifugation at successively higher speeds disrupts a cell and separation of its parts.

Cell Fusion

A new single cell is formed by a technique whereby cells or protoplasts are induced to fuse, often in pairs. Both precursor cells provide chromosomes. If the cells are genetically compatible, the nuclei may also fuse and a viable hybrid cell result, if the cells are genetically compatible. One or more chromosomes are expelled to form mininuclei in some cases, leaving a nucleus capable of division.

To produce a viable multicellular organism, it is possible to induce the development of such a product of cell fusion. Two gametes fuse to form a zygote in cell fusion that occurs naturally in sexual reproduction.

Cell Plate

At the midline of a dividing plant cell a membrane-bounded, flattened sac located inside which the new cell wall forms during cytokinesis.

Cellulose

Glucose residues comprises long unbranched chains by a high molecular weight polysaccharide (1,4)-linked. The cellulose chains form crystalline micelles separated by regions of randomized amorphous cellulose where cellulose is found in cell walls of higher plants and in some fungi as microfibrils.

Being freely available as a lignocellulose complex in wood, forest, and agricultural wastes, and in waste paper, cellulose is the most abundant carbohydrate.

Development of an effective, economic process for separation of cellulose from such materials and hydrolysis to glucose, in order to provide a low-cost fermentation substrate, development of an effective, economic process for separation of cellulose from such materials and hydrolysis to glucose, is a major challenge to biotechnologists.

Cell Wall

In the cells of plants, prokaryotes, fungi, and some protists there is a protective layer external to the plasma membrane. There are important structural components of cell walls, polysaccharides such as cellulose (in plants and some protists), chitin (in fungi), and peptidoglycan (in bacteria).

Cell-Mediated Immune Response

The activation of cytotoxic T cells, which defend against infected cells that involves the branch of adaptive immunity.

Cellular Component

A term coined by Gene Ontology to describe subcellular structures, locations and macromolecular complexes such as nucleus, telomere, and mitotic spindles.

Cellular Respiration

To produce ATP, the catabolic pathways of aerobic and anaerobic respiration, which break down organic molecules and use an electron transport chain.

Cellular Slime Mold

In the life cycle of a type of protist characterized by unicellular amoeboid cells and aggregated reproductive bodies.

Cellulose

Consisting of glucose monomers joined by β glycosidic linkages it includes a structural polysaccharide of plant cell walls.

Celsius Scale

Measures the freezing point of water at 0°C and the boiling point of water at 100°C there is a temperature scale (°C) equal to 5 /9(°F – 32).

Central Canal

In the center of the spinal cord that is continuous with the fluid-filled ventricles of the brain there is a narrow cavity.

Central Nervous System (CNS)

The brain and spinal cord where the integration occurs is the portion of the nervous system.

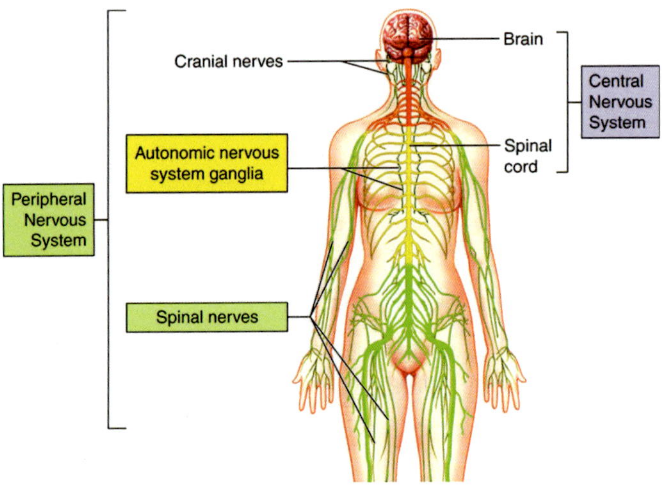

Figure 21. Image Showing CNS.

Source : Image By Neurotechedu

Central Vacuole

There is a large membranous sac with diverse roles in growth, storage, and sequestration of toxic substances in a mature cell.

Centriole

An animal cell composed of a cylinder of microtubule triplets arranged in a 9+0 pattern in the centrosome structure. A pair of centrioles consisted by a centrosome.

Centromere

The region on each sister chromatid where they are most closely attached to each other in a duplicate chromosome, by proteins that bind to specific DNA sequences; this close attachment causes a constriction in the condensed chromosome. (Identified by its DNA sequence chromosome has a single centromere that is uncondensed and unduplicated.)

Centrosome

During cell division there is an important structure present in the cytoplasm of animal cells that functions as a microtubule-organizing center. A centrosome has a pair of centrioles.

Cephalization

At the anterior end of the body, it involves an evolutionary trend toward the concentration of sensory equipment.

Cercozoan

Threadlike pseudopodia that feed an amoeboid or flagellated.

Cerebellum

Functioning in unconscious coordination of movement and balance depends on the part of the vertebrate hindbrain located dorsally.

Cerebral Cortex

The largest and most complex part of the mammalian brain is the surface of cerebrum, containing nerve cell bodies of the cerebrum; the part of the vertebrate brain most changed through evolution.

Cerebrospinal Fluid

The brain and the spinal cord is protected by blood-derived fluid that surrounds, protects against infection, nourishes, and cushioned it.

Cerebrum

The integrating center for memory, learning, emotions, and other highly complex functions of the central nervous system is controlled by the dorsal portion of the vertebrate forebrain, composed of right and left hemispheres.

Chaparral

Characterized by mild, rainy winters and long, hot, dry summers, a scrubland biome of dense, spiny evergreen shrubs found at midlatitudes along coasts where cold ocean currents circulate offshore.

Chaperonin

In the proper folding of other proteins assisted by a protein complex.

Character

Traits that may vary among individuals is an observable difference and inherent.

Character Displacement

In sympatric populations of two species is more divergent than in allopatric populations of the same two species.

Checkpoint

Stop and go-ahead signals can regulate the cycle by a control point in the cell cycle.

Chelicera

Characteristic of chelicerates has one of a pair of claws like feeding appendages

Chelicerate

A body divided into a cephalothorax and an abdomen and an arthropod that has chelicerae. Sea spiders, horseshoe crabs, scorpions, ticks, and spiders are some living chelicerates.

Chemical Bond

Sharing of outer-shell electrons or the presence of opposite charges on the atoms occurs because of an attraction between two atoms. Complete outer electron shells gained by the bonded atoms.

Chemical Energy

In a chemical reaction energy available in molecules for release; is a form of potential energy.

Chemical Equilibrium

In a chemical reaction, the state in which the rate of the forward reaction equals the rate of the reverse reaction in a chemical reaction. The reactants and products do not change with time of the relative concentrations.

Chemical Oxygen Demand (COD)

To oxidize organic matter (in a sample of polluted water, for example) using a chemical method, a measure of the amount of oxygen, expressed in milligrams per liter is required. The biochemical oxygen demand includes the chemical oxygen demands and is usually of a greater magnitude.

Chemical Reaction

Changes in the composition of matter leads to the making and breaking of chemical bonds.

Chemiosmosis

Energy stored in the form of a hydrogen ion gradient across a membrane to drive cellular work is used by an energy coupling mechanism, such as the synthesis of apt. Most apt synthesis in cells occurs by chemiosmosis, under aerobic conditions.

Chemoautotroph

An organism that obtains energy by oxidizing inorganic substances and needs only carbon dioxide as a carbon source.

Chemoheterotroph

Organisms that requires organic molecules for energy and carbon sources. sources.

Chemoreceptor

Responds to a chemical stimulus by a sensory receptor, such as a solute or an odorant.

Chemostat

Steady-state growth is maintained by providing the cells with a constant input of nutrients by a fermentation vessel or bioreactor. The level of a specific nutrient and the dilution rate (rate of removal of organisms from the vessel) matched to the cell growth rate can be regulated by the rate of growth.

Chiasma

In prophase I the X-shaped, microscopically visible region where crossing over has occurred earlier between homologous nonsister chromatids. With the two homologs remaining associated due to sister chromatid cohesion chiasmata become visible after synapsis ends.

Chimera

(1) A genetic make-up that is different from that of the surrounding cells of the same organism occurred by a segment of tissue (2) Tissues are of two different genotypes in an animal and plant. By grafting, by mixing cells of very early embryos or as a result of hybridization through normal crossing methods this can be achieved.

Chimeric Gene

By combining DNA sequences from several different sources an artificial gene is produced. For example, the sequence coding for an inducible enzyme can be spliced with the sequence for a small peptide. The hybrid protein can then be hydrolyzed to produce the required peptide product, if the chimeric gene is then inserted into a suitable host

Chi-Square Test

A given set of experimental values will be equaled or exceeded by chance alone by using a statistical test to determine the probability.

Chitin

In many fungal cell walls and in the exoskeletons of all arthropods, a structural polysaccharide, consisting of amino sugar monomers can be found.

Chloramphenicol

By binding onto the 50S subunits of bacterial and chloroplast ribosomes, an inhibitor of protein start acting and preventing peptide bond formation. To optimize replication of plasmids in cells of *E. coli* it is used.

Chlorohydrin

In the production of alkene oxides from alkenes is an intermediate chemical compound.

Chlorophyll

A molecule of the long-chain alcohol phytol complexed with magnesium containing a green pigment compound of a tetrapyrrole (porphyrin). In all autotrophic organisms, chlorophylls are the primary photosynthetic pigments are found.

Many distinct structural forms take place: chlorophyll A (in all oxygen producing photoautotrophs found); chlorophyll B (in higher plants and green algae found); chlorophylls C and D (in various algal groups are found); bacteriochlorophyll (in the green and purple bacteria are found). Based on absorption spectra various forms of chlorophyll A and B may be distinguished.

Isolated chlorophylls in organic solvents differ *in vivo*. Association of chlorophyll molecules, or chlorophyll dimers leads to the differences, with specific proteins and include the so-called reaction center complexes P700 and P680, which form the energy traps of photosystem I and photosystem II, respectively.

Chlorophyll A

Conversion of solar energy to chemical energy by a photosynthetic pigment that participates directly in the light reaction.

Chlorophyll B

Transformation of energy to chlorophyll A by an accessory photosynthetic pigment.

Chloroplast

A double membrane cell organelle found in cells of higher plants. The chloroplasts are green (due to chlorophyll) in higher plants, diameter is 3-10 μm and a biconvex lens shape. A complex lamellar structure, encloses the outer envelope, embedded in an amorphous protein stroma..

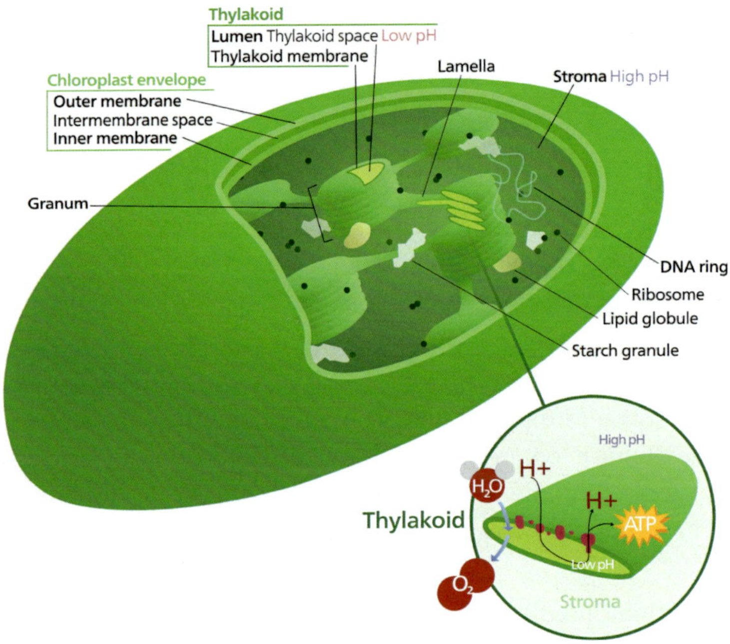

Figure 22. Image Showing Chloroplast.

Source : Image By Wikimedia Commons

Chlorosis

Due to a loss of chlorophyll, plant leaves become yellow. Due to nutrient deficiency, the effects of pests and disease or of chemicals, yellowing of leaves occurs (e.g., herbicides. fungicides or pesticides).

Choanocyte

A flagellated feeding cell found in sponges. Also called a collar cell, it has a collar-like ring that traps food particles around the base of its flagellum.

Cholesterol

An essential component of animal cell membranes formed by a steroid and acts as a precursor molecule such as many hormones, for the synthesis of other biologically important steroids.

Choline

A basic compound found as a constituent of phospholipids, a basic compound is found, such as lecithin, and the neurotransmitter acetylcholine.

Chondrichthyan

Member of the class Chondrichthyes, vertebrates with skeletons made mostly of cartilage, such as sharks and rays.

Chordate

Animals that at some point during their development have a notochord, member of the phylum Chordata, it also includes a dorsal, hollow nerve cord; pharyngeal slits or clefts; and a muscular, post-anal tail.

Chorionic Villus Sampling (CVS)

A small sample of the fetal portion of the placenta is removed for analysis in a technique associated with prenatal diagnosis, to detect certain genetic and congenital defects in the fetus.

Chromalveolata

In a current hypothesis of the evolutionary history of eukaryotes one of five supergroups of eukaryotes proposed. The alveolates and the stramenopiles chromalveolata's may have originated by secondary endosymbiosis and include two large protist clades.

Chromatid

From duplication during interphase resulted a longitudinal half chromosome. Meiotic chromatids may differ as a result of cross-over but mitotic chromatids are identical. During anaphase of mitosis (nonsexual division) chromatids of the same chromosome become separated and during anaphase II of meiosis (sexual division).

Chromatin

Eukaryotic chromosomes made up the complex of DNA and proteins. Chromatin exists in its dispersed form when the cell is not dividing, as a mass of very long, thin fibers that are not visible with a light microscope.

Chromatographic Detector

To monitor and quantify the products eluted from a chromatography column an instrument is used. Based on the physical, chemical, or biological properties of the molecules different techniques are used and to produce a colored or fluorescent complex involve prior conversion by chemical or enzymic mode.

The use of spectrophotometers. colorimeters, spectrophotofluorometers, refractometers Polarimeters etc., are involved in physical methods. Using enzyme-linked reactions specific for the required class of compound biological molecules may be monitored.

The fluid passes continuously from the column through the detector because such detectors often include a flow cell. For use with gas/liquid chromatographs several types of detectors have been developed specifically. Flame ionization and electron capture devices are included in this.

Chromatography

To separate or analyze mixtures of liquids, gases, compounds in solution or particles a technique is used. In terms of: the principles of separation (absorption. ion exchange, permeation, affinity) different types of chromatography are defined: the support material (silica. cellulose, starch, acrylamide or agarose gel, cellulose paper, cellulose acetate, etc.); the physical form of the matrix used (paper. thin layer. column).

Gas/liquid chromatography (GLC) and high performance (formerly high pressure) liquid chromatography (HPLC) are included in sophisticated automated chromatographic techniques. All forms of chromatography basic principles are similar.

The system includes three components: the sample; a solid matrix (stationary phase); a fluid (mobile phase). At the top of the column or at the edge of a gel slab, sheet of paper or thin layer support the mixture is placed. Through the stationary phase, the mobile phase is then passed.

At different rates by the mobile phase, the various components in the sample are carried through the stationary matrix, resulting in separation. Using column chromatography, the components are removed completely (eluted) and the mobile phase is then passed through a suitable detector where a response, proportional to the concentration of the various materials is used to establish the position and the quantity of the material.

The chromatograms are run with a suitable solvent using paper or thin layer chromatography, and for such a time that the compounds of interest are distributed across the sheet but are not eluted. By running such chromatograms twice at right-angles to each other in different solvents the resolution of a mixture may be increased. This is called as two-dimensional (2D) chromatography.

Based on their reaction with various chemical sprays the compounds are visualized and identified, once the matrix is dried, which may be combined with the use of ultraviolet light or fluorescence. The compounds may be located using autoradiography, where radioactive tracers have been used.

By comparing their behavior with standards (pure, authentic samples), an indication of the nature of such samples may be obtained. In terms of its RF the position of a given compound on a chromatograph may be defined. The ratio of the distance from the origin travelled by the compound to the distance moved by the solvent front is measured.

In terms of the ratio of the volume of liquid used to elute the sample to the volume of liquid used to elute the complete mixture characterized by the compounds in column system. The compounds may be characterized because of the concentration or pH at which they are eluted where the mobile phase consists of a solvent in which a gradual change in molar concentration or pH (a gradient) has been used.

Chromatography can indicate their chemical nature by comparison with standards and separates components of a mixture, subjecting them to further chemical and physical analysis the only way to prove conclusively the identity of the separated compounds.

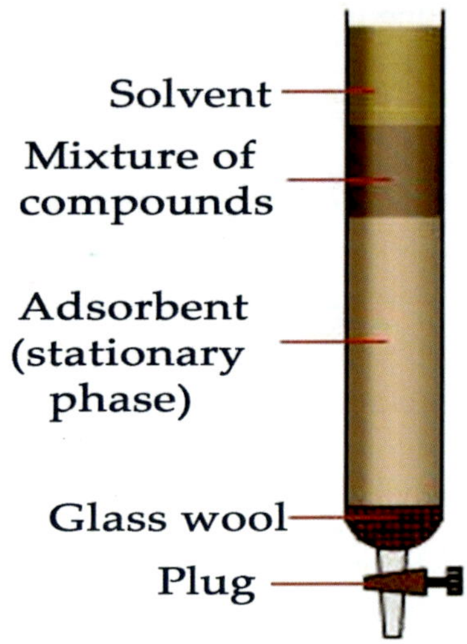

Figure 23. Column Chromatography.

Source : Image By Wikimedia Commons

Chromatophore

Colored pigments accumulated by a cell or organelle. The chromoplasts found in plants and the melanophores of animals included chromatophores. To describe the membrane vesicles associated with photosynthesis in prokaryotes the term is used.

Chromomere

From tightly coiled euchromatin, a small granule is formed, that stains deeply during prophase of mitosis and meiosis. During zygotene of meiosis the chromomeres of homologous chromosomes become aligned when chromosomes is paired.

Chromosome

The inheritable characteristics of an organism is carried by the DNA-carrying structure. In eukaryotes, the genetic material is contained within a membrane bounded nucleus, during interphase. When stained with basic dyes, are visible under the light microscope, this material becomes contracted and best visible during metaphase of cell division. Chromosomes are madeup of histone protein and DNA. The genes are linearly arranged on the chromosomes that are responsible for inheritance of traits. Each chromosome is consists of two chromatid that are joint together by central non staining region, the centromere. Terminal portion of chromatid is known as telomere that provides stability to the chromosomes. Although chromosomes may be present in the haploid, diploid or polyploid number, the number of chromosomes in a nucleus is constant for a given species. Pairs of homologous chromosomes in diploid cells are visually identical they differ from each other in length, shape, and position of the centromere.

The determination of the sex of the individual are associated with the distinguish of sex chromosomes and the rest termed autosomes.

The chromosome of prokaryotes is not associated with protein nor contained within a membrane and does not show secondary structure; instead, the genetic material generally consists of a circular molecule of DNA. A linear or circular portion of RNA consisted by the chromosomes in some viruses.

Chromosome Map

The relative positions of genes on chromosomes, plasmids, or viral genomes by a graphic representation. The physical structure based on microscopic examination (cytological map) indicated by such maps, the position of genes in terms of the nucleotide sequences or linkage units (linkage map).

Chromosome Mutation

It is also called an aberration or abnormality when a change in the structure or number of the chromosomes occurs.

Chromosome Polymorphism

For a given chromosome, the presence in a population of more than one gene sequence have.

Chromosomal Theory Of Inheritance

A basic principle in biology stating that the behavior of chromosomes during meiosis accounts for inheritance patterns and genes are located at specific positions (loci) on chromosomes.

Chylomicron

Fats mixed with cholesterol and coated with proteins is composed of a lipid transport globule.

Chyme

In the stomach the mixture of partially digested food and digestive juices are formed.

Chytrid

Mostly aquatic fungi with flagellated zoospores that represent an early diverging fungal lineage is the member of the fungal phylum Chytridiomycota.

Cilium

In eukaryotic cell a short appendage containing microtubules. Locomotion or moving fluid past the cell, a motile cilium is specialized for; a core of nine outer doublet microtubules and two inner single microtubules are used for its formation (the "9 + 2" arrangement) the plasma membrane in an extension ensheathed. A sensory and signaling role played by a primary cilium which is nonmotile; the two inner microtubules (the "9 + 0" arrangement) is lacked.

Circadian Rhythm

Even in the absence of external cues a physiological cycle of about 24 hours that persisted.

Cis-Trans Isomer

The same molecular formula and covalent bonds between atoms contained by one of several compounds, their atoms owing to the inflexibility of double bonds are differed in the spatial arrangements; formerly called a geometric isomer.

Citric Acid Cycle

The metabolic breakdown of glucose molecules begun in glycolysis by oxidizing acetyl CoA (derived from pyruvate), is completed in eight steps by a chemical cycle, to carbon dioxide; occurs within the mitochondrion in eukaryotic cells and in the cytosol of prokaryotes; the second major stage in cellular respiration, together with pyruvate oxidation.

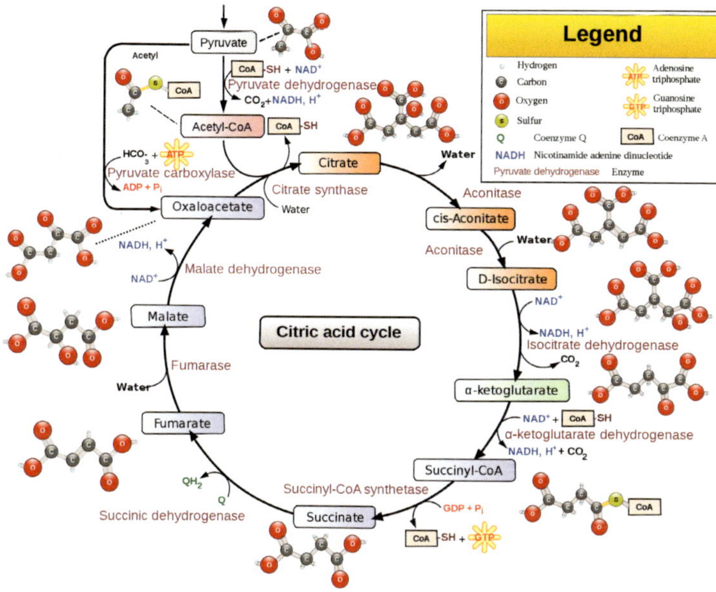

Figure 24. The Citric Acid Cycle.

Source : Image By Wikimedia Commons

Clade

All its descendants and an ancestral species is included by a group of species.

Cladistics

Organisms are placed into groups with systematic approach is called clades based primarily on common descent.

Class

The taxonomic category above the level of order, in Linnaean classification.

Cleavage

In animal cells, the process of cytokinesis, pinching of the plasma membrane is characterized. During early embryonic development, the succession of rapid cell divisions without significant growth, that converts the zygote to a ball of cells.

Cleavage Furrow

A shallow groove around the cell in the cell surface near the old metaphase plate in an animal cell is the first sign of cleavage.

Climate

At a given place prevailing weather conditions, in long term.

Climograph

In a particular region, a plot of the temperature and precipitation.

Cline

Along a geographic axis a graded change in a character.

Clitoris

During sexual intercourse, at the upper intersection of the labia minora that engorges with blood and becomes erect. It is an organ.

Cloaca

Found in many nonmammalian vertebrates but in few mammals a common opening for the digestive, urinary, and reproductive tracts.

Clonal Selection

An antigen selectively binds to and activates only those lymphocytes bearing receptors specific for the antigen involves in a process. For the stimulating antigen, the selected lymphocytes proliferate and differentiate into a clone of effector cells and a clone of memory cells specifics.

Clone

(1) Genetically identical individuals or cells in a lineage. (2) An individual that is genetically identical to another individual in a popular usage. (3) To make one or more genetic replicas of an individual or cell as a verb.

Cloning Vector

A DNA molecule that can carry foreign DNA into a host cell and replicate there, in genetic engineering. Plasmids and bacterial artificial chromosomes

(BACs) are included in a DNA molecule, which help in moving recombinant DNA from a test tube back into a cell, and viruses that involves in transferring recombinant DNA by infection.

Closed Circulatory System

Blood is confined to vessels and is kept separate from the interstitial fluid in a circulatory system.

Cluster

The grouping of similar objects in a multidimensional space. Clustering is used for constructing new features which are abstractions of the existing features of those objects. The quality of the clustering depends crucially on the distance metric in the space. In bioinformatics, clustering is performed on sequences, high - throughput expression (microarray data), and other experimental data.

Clusters of orthologous groups COG

NCBI compilations of evolutionarily related gene sequences from several microbial genomes. This site allows you to search by gene or cellular role and produce dendrograms to show sequence similarities.

Cnidocyte

The phylum Cnidaria; contains a capsule like organelle housing a coiled thread that is a unique specialized cell, explodes outward and functions in prey capture or defense when discharged.

Codominance

In the heterozygote, it is a situation in which the phenotypes of both alleles are exhibited because in separate, distinguishable ways both alleles affect the phenotype.

Codon

A particular amino acid or termination signal specified by a three-nucleotide sequence of DNA or mRNA; the genetic code is the basic unit.

Coding regions (CDS)

The portion of a genomic sequence bounded by start and stop codons that identifi es the sequence of the protein being coded for by a particular gene.

Coefficient of Relatedness

By two individuals the fraction of genes that, on average, are shared.

Coelom

Only from mesoderm a body cavity lined by tissue is derived.

Coelomate

A true coelom possessed by an animal (tissue completely derived from mesoderm a body cavity lined).

Coenocytic Fungus

It may contain hundreds or thousands of nuclei when a fungus that lacks septa and hence whose body is made up of a continuous cytoplasmic mass.

Coenzyme

Serving as a cofactor by an organic molecule. In metabolic reactions most vitamins function as coenzymes.

Coevolution

In response to selection imposed by the other of two interacting species by the joint evolution.

Cofactor

For the proper functioning of an enzyme, any nonprotein molecule or ion is required. Along with the substrate, during catalysis, cofactors can be permanently bound to the active site or may bind loosely and reversibly.

Cognition

Awareness, reasoning, recollection, and judgment included in the process of knowing.

Cognitive Map

In an animal's surroundings a neural representation of the abstract spatial relationships between objects.

Cohesion-Tension Hypothesis

The ascent of xylem sap by the leading explanation. It states that transpiration exerts pull on xylem sap, putting the sap under negative pressure or tension, and that the cohesion of water molecules transmits this pull along the entire length of the xylem from shoots to roots.

Collagen

The most abundant protein in the animal kingdom is glycoprotein in the extracellular matrix of animal cells that forms strong fibers, found extensively in connective tissue and bone.

Collecting Duct

Processed filtrate, called urine, is collected from the renal tubules in the kidney.

Collenchyma Cell

The plant without restraining growth by a flexible plant cell type that occurs in strands or cylinders that support young parts.

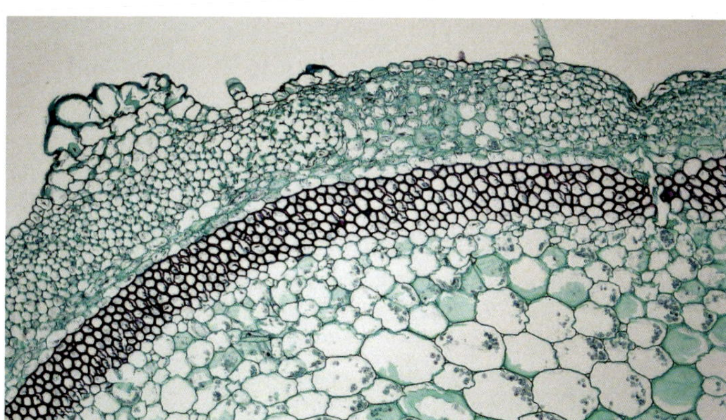

Figure 25. Herbaceous Dicot Stem: Ridge Collenchyma in Cucurbita.

Source : Image By Flickr

Colloid

Particles that remain suspended (because of their large size) rather than dissolved in that liquid is a mixture made up of a liquid.

Colon

In water absorption and formation of feces functions the largest section of the vertebrate large intestine.

Commensalism

One organism benefit but the other is neither helped nor harmed is a symbiotic relationship.

Interaction

A process involving transmission of, reception of, and response to signals, in animal behavior. In connection with other organisms and individual cells of multicellular organisms the term is also used.

Community

An assemblage of populations of different species living close enough together for potential interaction, all the organisms that inhabit a particular area.

Community Ecology

Interactions between species affect community structure and organization this study shows.

Companion Cell

By many plasmodesmata connected to a sieve-tube element that is a type of a plant cell and one or more adjacent sieve-tube elements is served by nucleus and ribosomes.

Competitive Exclusion

The concept for the same limited resources when populations of two similar species compete, one population will use the resources more efficiently that will eventually lead to the elimination of the other population and have a reproductive advantage.

Competitive Inhibitor

By entering the active site in place of the substrate the activity of an enzyme that reduced by a substance, whose structure it mimics.

Compiler

A computer program that translates a symbolic programming language into machine language so that the instructions can be executed by a computer.

Complement System

The inflammatory response amplified by a group of about 30 blood proteins, enhance phagocytosis, or directly lyse extracellular pathogens.

Complementary DNA (cDNA)

As a template and the enzymes reverse transcriptase and DNA polymerase *in vitro* using mRNA by a double-stranded DNA molecule. The exons of a gene corresponded by a cDNA molecule.

Complete Digestive Tract

An alimentary canal is known as a digestive tube that runs between a mouth and an anus.

Complete Dominance

The heterozygote and dominant homozygote are indistinguishable by phenotypes in this situation.

Complete Flower

Floral organs like sepals, petals, stamens, and carpels have in a flower.

Complete Metamorphosis

In the environment larva often functions very differently and transformation of a larva into an adult, then the larva.

Compound

In a fixed ratio a substance consisting of two or more different elements combined.

Compound Eye

In insects and crustaceans there is a type of multifaceted eye consisting of up to several thousand light-detecting, focusing ommatidia.

Concentration Gradient

A chemical substance increases or decreases in a region along with the density.

Condom

It fits over the penis to collect semen, made up of a thin latex rubber or natural membrane sheath.

Conduction

In direct contact with each other the transfer of thermal motion (heat) between molecules of objects occurred.

Conformer

In an environmental variable, an animal for which an internal condition conforms to (changes in accordance with) changes.

Conidium

During asexual reproduction a haploid spore produced at the tip of a specialized hypha in ascomycetes.

Conifer

Gymnosperm phylum largest member. Pines and firs are cone-bearing trees conifers.

Conjugation

Two cells that are temporarily joined the direct transfer of DNA between them in prokaryotes. Conjugation results in horizontal gene transfer when the two cells are members of different species. Two cells exchange haploid micronuclei but do not reproduce is asexual process in ciliates.

Connective Tissue

To bind and support other tissues animal tissue functions in animal tissue, an extracellular matrix scattered by a sparse population of cells scattered.

Conservation Biology

To sustain biological diversity at all levels, it is an integrated study of ecology, evolutionary biology, physiology, molecular biology, and genetics.

Conserved Domain (CD)

A domain that has been retained during evolution presumably due to its essential role within the protein's structure. Conserved domain searches are a part of the BLAST search.

Contractile Vacuole

Moving excess water out of certain freshwater protists helped by a membranous.

Control Element

By serving as a binding site for a transcription factor a segment of noncoding DNA that helps regulate transcription of a gene. In a eukaryotic gene's enhancer multiple control elements are presented.

Controlled Experiment

A control group that varies only in the factor being tested is compared in an experimental group in an experiment.

Convergent Extension

In such a way that the sheet of cells becomes narrower (converges) and longer (extends) in this process in which the cells of a tissue layer rearrange themselves.

Cooperativity

A shape change in one subunit of a protein caused by substrate binding is transmitted to all the other subunits by a kind of allosteric regulation, binding of additional substrate molecules to those subunits is facilitated.

Copepod

Important members of marine and freshwater plankton communities belong to any of a group of small crustaceans.

Coral Reef

Typically, a warm-water, tropical ecosystem dominated by the hard skeletal structures secreted primarily by corals. Some coral reefs also exist in cold, deep waters.

Figure 26. Coral Reef.

Source : Image By Pixabay

Corepressor

To bind to the operator and switch an operon off is allowed by a small molecule that binds to a bacterial repressor protein and changes the protein's shape.

Cork Cambium

The epidermis with thicker, tougher cork cells is replaced by cylinder of meristematic tissue in woody plants.

Corpus Callosum

To process information together the right and left cerebral hemispheres are connected in mammals, enabling the hemispheres by a thick band of nerve fibers.

Corpus Luteum

From the collapsed follicle after ovulation and produces progesterone that helps in the formation of a secreting tissue in the ovary.

Cortex

(1) Due to the presence of multiple microfilaments the outer region of cytoplasm in a eukaryotic cell, lying just under the plasma membrane, that has a more gel-like consistency than the inner regions. (2) In a root or eudicot stem, ground tissue that is between the vascular tissue and dermal tissue in plants.

Cortical Nephron

In the renal cortex a nephron with a loop of Henle located, in the mammals.

Cotransport

Transport of another against its own concentration gradient the coupling of the "downhill" diffusion of one substance to the "uphill".

Cotyledon

An angiosperm embryo of a seed leaf. Some species have one cotyledon, others have two.

Countercurrent Exchange

In the opposite directions exchange of a substance or heat between two fluids flowing. For example, water passing over the gill, blood in a fish gill flows in the opposite direction, maximizing diffusion of carbon dioxide out of the blood and oxygen into.

Countercurrent Multiplier System

To facilitate exchange of materials and generate concentration gradients, a countercurrent system in which energy is expended in active transport

Cassulacean Acid Metabolism (CAM)

In the family Crassulaceae, discovered first, an adaptation for photosynthesis in arid conditions. A plant takes up CO2 and incorporates it into a variety of organic acids at night in this process, CO2 is released from organic acids, during the day, for use in the Calvin cycle.

Crista

The inner membrane of a mitochondrion is an infoldings. The synthesis of ATP (ATP synthase) electron transport chains and molecules of the enzymes catalyzing in the inner membrane houses.

Critical Load

Plants without damaging ecosystem integrity absorbs the amount of added nutrient, usually nitrogen or phosphorus.

Crop Rotation

To restore concentrations of fixed nitrogen in the soil, the practice of planting nonlegumes one year and legumes in alternating years.

Cross-Fostering Study

In the care of adults, the young of one species are placed from another species in behavioral study.

Cross-Pollination

In angiosperms, the transfer of pollen from an anther of a flower on one plant to the stigma of a flower on another plant of the same species.

Crustacean

Lobsters, crayfishes, crabs, shrimps, and barnacles included in a member of a subphylum of mostly aquatic arthropods.

Culture

The behavior of individuals in a population is influenced by the system of information transfer through social learning or teaching.

Cuticle

(1) In terrestrial plants a waxy covering on the surface of stems and leaves that prevents desiccation. (2) For different functions layers of protein and chitin that are variously modified and consisted by the exoskeleton. (3) The body of a nematode covered by the tough coat.

Cyclic AMP (cAMP)

A common intracellular signaling molecule (second messenger), Cyclic adenosine monophosphate a ring-shaped molecule made from ATP, in eukaryotic cells. Some bacterial operons also regulated by it.

Cyclic Electron Flow

The light reactions of photosynthesis that involves only photosystem I responsible for the flow of electron in a route and that produces ATP but not NADPH or O2.

Cyclin

In a cyclically fluctuating concentration occurred by a cellular protein and in regulating the cell cycle played a vital role.

Cystic Fibrosis

For a chloride channel protein caused by a recessive allele in a human genetic disorder; by an excessive secretion of mucus and consequent vulnerability to infection is characterized; and it is fatal if untreated.

Cytochrome

A component of electron transport chains containing iron protein in the mitochondria and the plasma membranes of prokaryotic cells and chloroplasts of eukaryotic cells.

Cytogenetic Map

Chromosomal features are distinguishable in a microscope, a map of a chromosome that locates genes with respect to.

Cytogenetics

Chromosome form and function linked by a branch of biology. For the diagnosis of inherited disease, it may be used as a means.

Cytokine

Any of a group of small proteins secreted by several cell types, including macrophages and helper T cells, that regulate the function of other cells.

Cytokinesis

To form two separate daughter cells immediately after mitosis, the division of the cytoplasm takes place, meiosis I, or meiosis II.

Cytokinin

To stimulate cell division, any of a class of related plant hormones that retard aging and act in concert with auxin, the pathway of differentiation, and control apical dominance is influenced.

Cytoplasmic Determinant

The course of early development, influenced by a maternal substance, such as a protein or RNA, that when placed into an egg through regulating the expression of genes that affect the developmental fate of cells.

Cytoplasmic Inheritance

Inheritance other than the main chromosomal DNA of prokaryotes, or that associated with the nuclear DNA in eukaryotes. In eukaryotes, cytoplasmic inheritance is mainly associated with the transmission of mitochondria is mainly associated in eukaryotes and/ or chloroplasts from one maternal generation to the next. In prokaryotes, due to the transmission of plasmids from one generation to the next in prokaryotes it is occurred.

Cytoplasmic Streaming

Involving interactions of myosin and actin filaments, a circular flow of cytoplasm that speeds the distribution of materials within cells.

Cytoskeleton

Throughout the cytoplasm filaments are extended by a network of microtubules, microfilaments, and intermediate and a variety of mechanical, transport, and signaling functions are served.

Cytosine (C)

Nucleic acids and nucleotides are the derivatives constitutents of a pyrimidine base.

Cytotoxic Drug

It is kind of drug that kills cells. To refer to compounds that kill cancer cells, the term is sometimes used specifically.

Cytotoxic T Cell

It kills infected cells as well as certain cancer cells and transplanted cells, when activated.

D

Dalton

The same as the atomic mass unit, or amu is. a measure of mass for atoms and subatomic particles.

Dark Reaction

Through the photosynthetic carbon reduction cycle (PCR) in the carbon assimilation and associated pathways, including the c4 cycle and c2 photorespiratory cycle. Soluble enzymes found in the stroma of the chloroplasts catalyzed the dark reactions and use ATP and NADH generated in the light reactions of photosynthesis as a source of energy and reducing power.

Darwinian Fitness

Relative contribution to the following generations is determined by the relative fitness of one genotype compared with another.

Data Logger

An electrical signal accepted by a device, which either be digital or analog, and a numerical output is converted by this.

Data Mining

The ability to query very large databases in order to satisfy a hypothesis (" top - down " data mining); or to interrogate a database in order to generate new hypotheses based on rigorous statistical correlations (" bottom - up " data mining).

Data Processing

The systematic performance on data of such operations as handling, merging, sorting, and computing. The semantic content of the original data should not be changed, but the semantic content of the processed data may be changed.

Data Warehouses

Vast arrays of heterogeneous (biological) data, stored within a single logical data repository, that are accessible to different querying and manipulation methods.

Day-Neutral Plant

The length of day does not influence the tendency to flower in a plant (the plant remains in a condition of total darkness depends on the length of plant).

Deaminase

An enzyme that catalyzes the removal of an amino (-NH2) group from a chemical compound (usually an amino acid or amine).

Debranching Enzyme

An enzyme that hydrolyzes the (1 ,6Hinks) found in amylopectin and similar polysaccharides.

Decant

The breakup of a solid from a liquid by pouring off the liquid from a settled solid; pouring a fluid from one container to another is a process.

Decapod

Lobsters, crayfishes, crabs, and shrimps are the group of crustaceans of a member.

Decomposer

Absorption of nutrients from nonliving organic material of an organism such as corpses, fallen plant material, and the wastes of living organisms and converts them to inorganic forms; is a detritivore.

Decontamination

Enabling an area to be entered by the removal of a hazardous material such as a radioisotope or a pathogenic organism, or the object to be reused, without having to resort protective clothing or take precautions.

Deductive Reasoning

From a general premise, specific results are predicted by a type of logic.

Deep-Sea Hydrothermal Vent

Association of volcanic activity on or near the seafloor a dark, hot, oxygen-deficient environment. The producers are chemoautotrophic prokaryotes in a vent community.

De-Etiolation

The changes a plant shoot undergoes in response to sunlight when a plant shoot undergoes; also known informally as greening.

Deficiency Disease

A diet that lacks an essential nutrient results a disease, such as a vitamin, essential amino acid, or trace element.

Degenerate Codons

In the third letter variations nucleotide triplets which although containing, with the same initial letters, will code for the same amino acid as others.

Dehydration

The removal or drying of water. For examination under a microscope, it is used as a means of preserving biological material or preparing specimens. To preserve bacteria, yeasts and fungi, dehydration, using careful freeze-drying techniques is used, in culture collections.

Dehydration Reaction

Two molecules become covalently bonded to each other with the removal of a water molecule in a chemical reaction.

Dehydrogenase

A type of oxidoreductase, by transferring hydrogen from one compound to another by an enzyme that catalyzes oxidation-reduction reactions.

Deionization

The separation of ions from a solution. Using an ion exchange resin, or a mixture of such resins, it is usually achieved.

Deionized Water

Using ion exchange resins, a pure water of low conductivity from which cations and anions have been removed. It is usual to distil the water prior to treatment with resin if very pure deionized water is required.

Deletion

(1) Resulting from the loss of a fragment through breakage because of deficiency in a chromosome. (2) During nuclear division, a chromosome aberration in which a section is lost. Since a loop is formed to allow normal pairing on each side of the deleted region deletions can be recognized at prophase in meiosis, in some cases.

Demographic Transition

A shift from high birth and death rates to low birth and death rates, in a stable population.

Demography

In the vital statistics of populations, especially birth rates and death rates, the study of changes over time.

Denaturation

Due to the loss of non-covalent bonding a change in conformation and a loss of secondary and tertiary structure that occurs when proteins and nucleic acids are exposed to extreme temperature or pH, non-physiological concentrations of salt, detergents, organic solvents, or other chemical agents.

It can lead to dissociation of the helical structure in nucleic acids and unfolding of the polypeptide chain in proteins. Changes in solubility, physical properties and biological activity may be associated with the changes. Changes may be reversible.

Denatured DNA

By breaking the hydrogen bonds joining the two complementary strands of DNA that has been converted from a double-stranded to a single-stranded form.

Denatured Protein

By exposure to a destabilizing agent such as heat, a protein that has lost its natural configuration.

Dendrite

The cell body of a neuron and makes synaptic contacts with other neurons raised from a short process; During the transfer of a nerve impulse the input region for excitation from one neuron to the next.

Dendritic Cell

Thereby initiating a primary immune response, an antigen-presenting cell, located mainly in lymphatic tissues and skin, in presenting antigens to help T cells, that is particularly efficient.

Figure 27. Ultrastructural Characteristics of Mature And Apoptotic/Necrotic Myeloid DCs Investigated By SEM.

Source : Image By Wikimedia Commons

Dendrogram

Another name for phylogenetic tree. The clustalw guide tree is often referred to as a dendrogram, it is a file with a .dnd extension.

Denitrification

By facultative aerobic soil bacteria under anaerobic condition catalyzed by the reduction of nitrates to nitrites, nitrous oxide or nitrogen catalyzed.

Densitometer

To evaluate the number of various compounds, present on the plate or gels a device is used to scan chromatographic plates or electrophoretic gels. Alight source, a sample holder, and a detector (photomultiplier) are the essential parts consisted.

Through the light (visible or ultraviolet) beam and the changes in density, reflected light or fluorescence (depending on the mode of operation) the sample is passed continually, are recorded on a chart, or using a data logger. Using a suitable dye prior to analysis, by the samples being scanned are often are stained.

Density

Per unit area or volume, the number of individuals.

Deoxyribonucleic Acid (DNA)

Usually a double-stranded helix, in which each polynucleotide strand consists of nucleotide monomers, a nucleic acid molecule with a deoxyribose sugar and the nitrogenous bases adenine (A), cytosine (C), guanine (G), and thymine (T); capable of being replicated and determining the inherited structure of a cell's proteins.

Deoxyribose

The sugar component of DNA nucleotides, having one fewer hydroxyl group than ribose, the sugar component of RNA nucleotides.

Depolarization

A change in a cell's membrane potential such that the inside of the membrane is made less negative relative to the outside. For example, a neuron membrane is depolarized if a stimulus decreases its voltage from the resting potential of 70 mV in the direction of zero voltage.

Desert

By very low precipitation, a terrestrial biome characterized.

Desmosome

Cells those functions as a rivet, fastening cells together, as a type of intercellular junction in animal.

Determinate Cleavage

The developmental fate of each embryonic cell very early is a type of embryonic development in protostomes that rigidly casts.

Determinate Growth

Growth stops after a certain size is reached.

Detritivore

A consumer that derives its energy and nutrients from nonliving organic material such as corpses, fallen plant material, and the wastes of living organisms; a decomposer.

Deuteromycete

With no known sexual stage, traditional classification for a fungus.

Deuterostome Development

By the development of the anus from the blastopore is distinguished by a development mode, in animals; by the body cavity forming as out pockets of mesodermal tissue and often also characterized by radial cleavage.

Development

In an organism's changing gradually from a simple to a more complex or specialized form, some events are involved.

Diabetes Mellitus

An endocrine disorder marked by an inability to maintain glucose homeostasis. Type I diabetes results from autoimmune destruction of insulin-secreting cells; daily insulin injections required for the treatment. Type-2 diabetes is the result of reduced responsiveness of target cells to insulin; lack of exercise and obesity are risk factors.

Diacylglycerol (DAG)

In the plasma membrane, second messenger produced by the cleavage of the phospholipid PIP2.

Diagnosis

Generally, any scientific procedure that allows a precise result or classification to be reached involved, the process of determining by examination the nature and circumstances of a diseased condition; the conclusions reached from such an examination.

Diagnostic

A symptom or characteristic having value in diagnosis, descriptive of a diagnosis. The term is used to describe a test or procedure that may be either qualitative or quantitative in health care and experimental biology, and to reveal the presence or number of specific substances (such as metabolites, enzymes, or antibiotics) it is designed, thus indicating a disease or pathological condition.

Diagnostic Enzyme

In order to obtain a qualitative or quantitative indication of the presence of a given compound an enzyme used as a specific diagnostic agent, which serves as the enzyme substrate.

Diagnostic Kit

To carry out one or more specific assays, a package that contains all the required reagents, in suitable form, is required, which may be based on immune, chemical, or enzyme-catalyzed reactions. By health authorities, to be accepted for use in health care, such reagents and procedures require acceptance and validation.

Diakinesis

The centromeres move further apart, a stage in the prophase of the first division of meiosis and the double chromosomes take on the appearance of rings. The nucleolus begins to disappear after the chromosomes continue to contract.

Dialysis

Through a selectively permeable membrane the separation of substances of different molecular weight by diffusion. To remove low molecular weight compounds dialysis is used, such as salts from solutions of high molecular weight compounds such as proteins.

In a sealed semipermeable membrane, the mixed solution is placed, which is immersed in water or a suitable buffer. The high molecular weight materials are retained whereas low molecular weight compounds diffuse through the membrane.

Diaphragm

(1) The bottom wall of the thoracic cavity, a sheet of muscle that forms in mammals. The diaphragm pulls air into the lungs while contraction. (2) Before sexual intercourse, a dome-shaped rubber cup fitted into the upper portion of the vagina. It blocks the passage of sperm into the uterus.

Diapsid

A member of the amniote clade differentiated by two openings on either side of the skull. Lepidosaurs and archosaurs are examples of diapsids.

Diastole

In the cardiac cycle, the stage where a heart chamber relaxes and fills with blood.

Dicot

Flowering plants with two embryonic seed leaves, or cotyledons. According to recent molecular data dicots do not form a clade. Species originally classed as dicots are now divided into eudicots, magnoliids, and many lineages of basal angiosperms.

Differential Gene Expression

The expression of various sets of genes by cells with the same genome.

Differentiation

The process by which a cell or a group of cells become structurally and functionally specialised.

Diffusion

The spontaneous motion of a substance down its concentration or electrochemical gradient, from a more concentrated region to a less concentrated region.

Digestion

The breaking down of food into molecules small enough for the body to absorb is the second stage of food processing in animals.

Dihybrid

An organism having two genes of interest that are heterozygous. Dihybrids are the progeny of parents who are doubly homozygous for distinct alleles. Parents with genotypes AABB and aabb, for example, generate a dihybrid of genotype AaBb.

Dihybrid Cross

A cross between two organisms that are heterozygous for both of the traits being studied is called a dihybrid cross (or the self-pollination of a plant that is heterozygous for both characters).

Dikaryotic

A fungal mycelium having two haploid nuclei, one from each parent, per cell.

Dinoflagellate

A member of a generally unicellular photosynthetic group of algae having two flagella located in perpendicular grooves in cellulose plates covering the cell.

Dinosaur

A member of a highly diverse group of reptiles having different body shapes, sizes, and habitats. The only extant dinosaurs are birds.

Figure 28. Image Showing T-Rex Dinosaur.

Source : Image By Maxpixel

Dioecious

Having male and female reproductive organs on distinct individuals of the same species in plant biology.

Diploid Cell

A cell with two sets of chromosomes (2n), a single set inherited from each parent.

Diplomonad

A protist with redesigned mitochondria, two nuclei of equal size, and multiple flagella.

Directional Selection

Natural selection wherein individuals at one extreme of the phenotypic range that live or reproduce more successfully than others.

Disaccharide

A glycosidic linkage generated by a dehydration event, connecting two monosaccharides to form a double sugar.

Dispersal

Individuals or gametes moving away from their parental location. A population or species' geographic range may be expanded as a result of this movement.

Dispersion

The pattern of spacing between individuals inside a population's boundaries.

Disruptive Selection

A case of natural selection where individuals at both ends of a phenotypic range survive and reproduce more successfully than those with intermediate phenotypes.

Distal Tubule

The portion of a nephron, in the vertebrate kidney, that aids in refining filtrate and empties it into a collecting duct.

Distance matrix

Matrix that contains all the pairwise distances between sequences in a data set. Distance matrices safe often used to compute phylogenetic trees. Not a substitution matrix.

Disturbance

An event which is natural or human-caused, that alters a biological community and generally removes organisms from it. Fires and storms, are some disturbances that play a vital role in structuring various communities.

Disulfide Bridge

The strong covalent bond produced when the sulphur of one cysteine monomer links to the sulphur of another cysteine monomer.

DNA (Deoxyribonucleic Acid)

A double-stranded, helical nucleic acid molecule, formed by nucleotide monomers with a deoxyribose sugar and the nitrogenous bases adenine (A), cytosine (C), guanine (G), and thymine (T); which can be duplicated and dictate the hereditary structure of a cell's proteins.

DNA Ligase

The linking enzyme required for DNA replication; catalyzes the covalent bonding of the 3 end of one DNA fragment (such as an Okazaki fragment) to the 5 end of another DNA fragment (such as a growing DNA chain).

DNA Methylation

The presence of methyl groups on the DNA bases of plants, animals, and fungi (typically cytosine). (It is also referred as the process of adding methyl groups to DNA bases sometimes.)

DNA Microarrays

The deposition of oligonucleotides or cDNAs onto an inert substrate such as glass or silicon. Thousands of molecules may be organized spatially into a high - density matrix. These DNA chips may be probed to allow expression monitoring of many thousands of genes simultaneously. Uses include study of polymorphisms in genes, de novo sequencing, or molecular diagnosis of disease.

DNA Microarray Assay

A technique for simultaneously detecting and measuring the expression of thousands of genes. A large number of single-stranded DNA fragments representing different genes are attached to a glass slide and evaluated for hybridization with labelled cDNA samples.

DNA Polymerase

The enzyme catalyzing the addition of nucleotides to the 3' end of an existing chain to lengthen new DNA (for example, at a replication fork). In E. coli, DNA polymerase III and DNA polymerase I are various different DNA polymerases that play an important role in DNA replication.

DNA Replication

Process of copying a DNA molecule; also called DNA synthesis.

Domain

A taxonomic level higher than the kingdom. Archaea, Bacteria, and Eukarya are the three domains. (2) A protein's distinct structural and functional area.

Dominant Allele

A completely expressed allele in the phenotype of a heterozygote.

Dominant Species

A species in a community with significantly greater concentrations or biomass than other species. Dominant species have a strong influence on the distribution and occurrence of other species.

Dopamine

A catecholamine neurotransmitter, like epinephrine and norepinephrine.

Dormancy

A condition characterised by a slow metabolic rate and a lack of growth and development.

Dorsal

With radial or bilateral symmetry, referring to the top of an animal.

Dorsal Lip

The area on the dorsal side of the amphibian embryo above the blastopore.

Double Bond

A double covalent bond with two atoms that share two pairs of valence electrons.

Double Circulation

The circulatory system which is divided into pulmonary and systemic circuits, with blood passing through the heart when each circuit is completed.

Double Fertilization

The process of fertilization where two sperm cells combine with two cells in the female gametophyte (embryo sac) to generate the zygote and endosperm in angiosperms.

Double Helix

Native DNA in the spiral structure, with two contiguous antiparallel polynucleotide strands coiled around an imaginary axis.

Down Syndrome

A genetic condition in humans resulting from the additional presence of chromosome 21. It is marked by developmental delays, heart and other problems, and is normally curable or non-life threatening.

Duchenne Muscular Dystrophy

A type of muscular dystrophy caused by a sex-linked recessive allele. This human genetic disorder further results in gradual weakness and muscular tissue loss.

Duodenum

The first segment of the small intestine, where stomach chyme interacts with digestive secretions from the pancreas, liver, and gallbladder, as well as from intestinal wall gland cells.

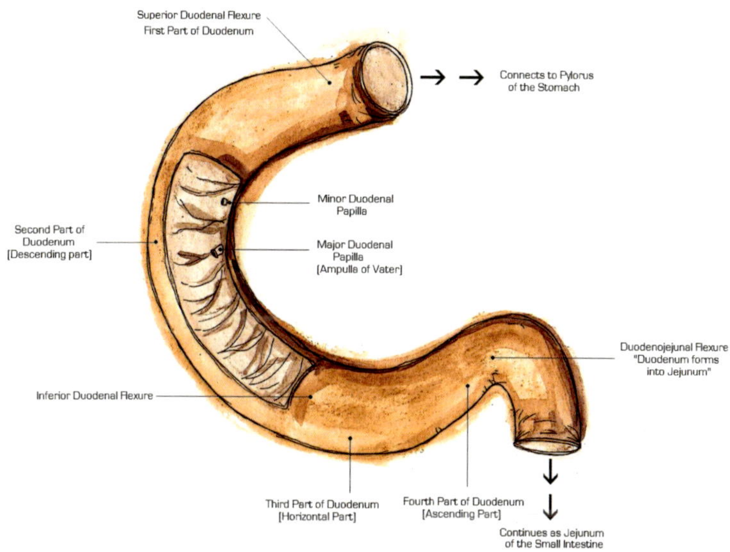

Figure 29. *Anatomical Illustration Using Traditional Media To Display The Duodenum.*

Source : Image By Wikimedia Commons

Duplication

A chromosome structural abnormality caused by fusion with a homologous chromosome fragment, resulting in chromosome duplication.

Dynamic programming

A program that allows a computer to explore efficiently all possible solutions to certain types of complex problems; it divides a problem into reasonably sized subproblems and uses parts to compute the final answer.

Dynamic Stability Hypothesis

The idea that long food chains are less stable than short ones.

Dynein

A big motor protein found in cilia and flagella that extends from one microtubule doublet to the next. ATP hydrolysis causes changes in dynein shape, which causes cilia and flagella to bend. It is characterised by increasing muscular weakness and tissue loss.

E

E Site

During translation, one of the three binding sites for tRNA on a ribosome. The E site is where tRNAs are ejected from the ribosome. (The letter E stands for exit.)

Ecdysis

When the hard exoskeleton is shed on a regular basis. Arthropods go through this process, which allows them to expand even more.

Ecdysozoan

A member of a group of animal phyla as identified by molecular evidence as a clade. Many ecdysozoans are moulting animals.

Ecdysteroid

Secreted by the prothoracic glands it is a steroid hormone, that triggers molting in arthropods.

Echinoderm

A marine deuterostome that is slow-moving or sessile, having a water vascular system and, in larvae, bilateral symmetry. Sea stars, brittle stars, sea urchins, feather stars, and sea cucumbers are some of the Echinoderms.

Ecological Footprint

A person, city, or nation's total land and water area required to produce all of the resources it consumes and absorb all of the waste it produces.

Ecological Niche (Nich)

The totality of a species' usage of biotic and abiotic resources in its habitat.

Ecological Species Concept

With respect to an ecological niche, the sum of how members of a species interact with non- living and living aspects of their environment.

Ecological Succession

When a community that emerges in an otherwise lifeless environment, after a community's species makeup changes due to a disruption.

Ecology

The science of interaction of organisms with each other and their surroundings.

Ecosystem

A unit made up of all the living species in a community, as well as the environment in which they live, particularly its chemical and physical characteristics. It is the basic unit of ecological research.

At different trophic levels, the creatures are classified as primary producers, herbivores, carnivores, and decomposers. The characterisitics of the abiotic component of the ecosystem determines the type of organisms found in a given area.

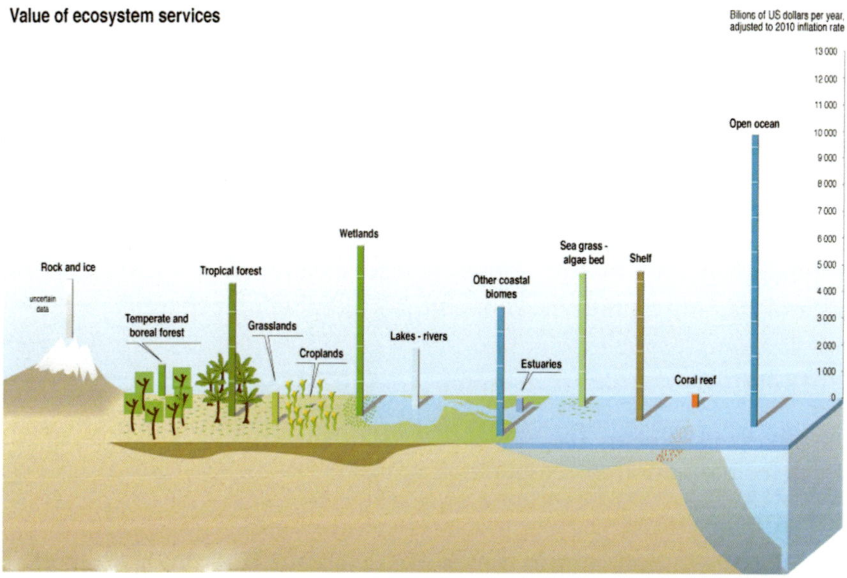

Figure 30. Value Of Ecosystem Services.

Source : Image By Flickr

Ecosystem Ecology

The study of energy flow and chemical cycling in an ecosystem's diverse biotic and abiotic components.

Ecosystem Engineer

An organism that causes physical changes in the environment to alter community organisation.

Ecosystem Service

An ecosystem's function that benefits humans either directly or indirectly.

Ecotone

The transformation from one type of habitat or ecosystem to another, such as from a forest to a grassland, is known as an ecotone.

Ectomycorrhiza

The association of a fungus with a plant root system where the former surrounds the roots but does not cause invagination of the host (plant) cells' plasma membranes

Ectomycorrhizal Fungus

A symbiotic fungus that develops hyphae sheaths on the surface of plant roots and also grows into the root cortex's extracellular spaces.

Ectothermic

Referring to organisms for which majority of heat is provided by external sources for regulation of temperature.

Ectotrophic Mycorrhiza

A basidiomycete fungus that grows in association with its host plant. The Hartig net is formed when fungal hyphae enter between the cells of the cortex to form a meshwork. The link causes root hair production at the root tips to be suppressed, as well as other morphological alterations in the roots.

The fungus aids in the mobilisation of phosphate and other ions from the soil to the host plant, which provides the fungus with glucose and other nutrients. This type of mycorrhizal relationship may be necessary for the plant to grow, particularly in poor soil.

Edaphic Factor

A physical, chemical, or biological property of soil that has an impact on plant growth.

Ediacaran Biota

An early group of multicellular, soft-bodied eukaryotes discovered in fossils dating from 565 million to 550 million years ago.

Edman Chemistry

A sequence of chemical reactions typically used as the basis for spinning cup, solid phase, gas phase and micro-sequencing of proteins. Each cycle of reactions results in the removal of one amino acid from a peptide's N-terminal end and the creation of a new peptide that is one amino acid shorter.

Two steps make up the entire reaction. Phenyl-isothiocyanate is coupled to the amino end of the peptide under basic circumstances to generate phenyl-thiocarbamyl peptide in the first coupling step.

The derivatized amino acid is removed as its anilinothiazoline in the second cleavage reaction by treatment with strong anhydrous acid. For further investigation, the phenyl-thiohydantoin (PTH) amino acid derivative is commonly converted.

Effective Population Size

An estimation of a population's size based on the number of successful breeding females and males; usually smaller than the overall population.

Effector Cell

(1) A muscle cell or gland cell that governs the body's reaction to stimuli in response to messages from the brain or other nervous system processing centres. (2) A lymphocyte capable of mediating an adaptive immune response that has undergone clonal selection.

Efferent

A nerve or neuron that transmits data from the central nervous system to the peripheral effectors.

Effluent

The mixed liquor discharged from a reactor. Specifically the term may be used to refer to a liquid, with or without suspended particles, or a solid discharge from a reactor or process plant that does not represent the required product and is thus regarded as a waste and will require further treatment or will cause pollution problems.

Egg-Polarity Gene

A maternal effect gene which is a gene that helps influence the orientation (polarity) of the egg.

Ejaculation

Sperm propulsion through the muscular vas deferens, ejaculatory duct, and urethra from the epididymis.

Ejaculatory Duct

The vas deferens and a duct from the seminal vesicle meet to create a small piece of the ejaculatory pathway in mammals. The ejaculatory duct transports sperm from the vas deferens to the urethra.

Electrocardiogram (ECG Or EKG)

The record of the electrical impulses traveling through heart muscle during the cardiac cycle.

Electrochemical Gradient

The concentration differential of an ion across a membrane (a chemical force) and the ion's tendency to migrate relative to the membrane potential determine the diffusion gradient of an ion (an electrical force).

Electrochemical Sensor

An analytical device based on either a potentiometric or an ampere-metric electrode. The response (in millivolts) of a potentiometric electrode is a logarithmic function of the concentration. Ion-selective electrodes are built on this foundation.

The response (in microamps) of an ampere-metric sensor is a linear function of concentration. The current recorded is caused by the species being measured diffusing towards the electrode as it is consumed at the electrode surface. This is how the oxygen electrode works.

Electrodialysis

A technique used for the deionization of solutions. A voltage is delivered across ionspecific semipermeable membranes that separate the sample from the rest of the compartment.

Larger molecules or uncharged chemicals are held within the chamber, while charged ions migrate to the appropriate electrode. Desalination of sea water is done on a big scale using electrodialysis. It's also utilised to deionize whey permeates, cell hydrolyzates, and other liquids.

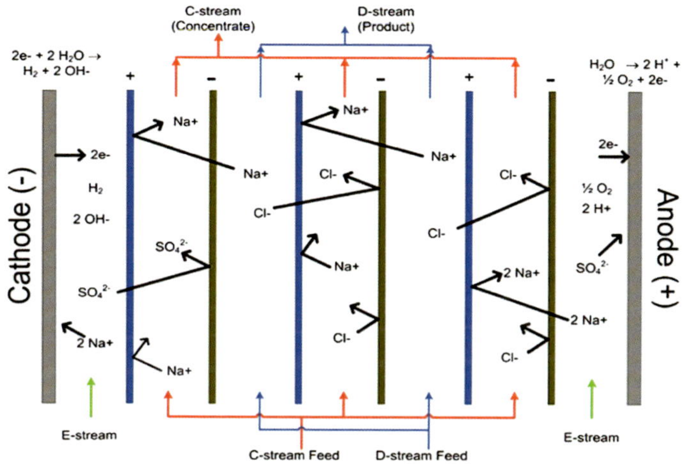

Figure 31. Electrodialysis.

Source : Image By Wikimedia Commons

Electrofusion

A technique for causing animal cells and protoplasts to fuse. Exposure to a low-level nonuniform electric field aligns the cells. Di-electrophoresis is a mechanism that creates dipoles within cells. The orientation of the cells is determined by the attractive forces between the dipoles.

A microsecond high-voltage discharge is then applied to the cells. This causes minute holes in cell membranes where cells come into contact. The gaps in the cells expand, allowing the contents of the cells to mix and some cells to fuse together.

Electrogenic Pump

An active transport protein that pumps ions across a membrane while generating voltage.

Electrolysis

The electrical decomposition of a substance or the destruction of tumors, hair roots, and other structures in a similar manner.

Electrolyte

(1) A conducting medium in which electrons and matter flow in the same direction. (2) A material that dissociates into ions when dissolved in water, forming an electrical conductor.

Electromagnetic Receptor

A receptor of electromagnetic energy, like the visible light, electricity, or magnetism.

Electromagnetic Spectrum

The complete electromagnetic spectrum, with wavelengths ranging from less than a nanometer to more than a kilometre.

Electron

A single-negative electrical charge subatomic particle having a mass of about 1/2,000 that of a neutron or proton. An atom's nucleus has one or more electrons that travel around it.

Electron Microscope

An equipment for observing and taking photographs of magnified objects. The electron microscope has a far better degree of resolution than the light microscope because the wavelength of the electrons utilised as a radiation source is much shorter than that of light.

To avoid electron scatter from collisions with air molecules, the system must be operated at a high vacuum. The electron beam interacts with a fluorescent screen or photographic film to form an electron micrograph, which visualises the image. The sort of sample utilised varies depending on the electron microscope.

The electron beam is sent through the specimen in the transmission microscope, which has been prepared in the shape of an ultrathin section in the case of bigger organisms. These slices are glutaraldehyde-fixed, dehydrated, and stained with electron-dense compounds like uranyl acetate, osmium tetroxide, lead citrate, and so on.

Viruses, cell fragments, and nucleic acids can be put intact on a grid and shaded with metal (e.g., platinum or gold) or negatively stained with another electron-dense material. Scanning microscopy samples can be much larger.

The sample may be coated with electron-dense substance or the preparation may be dehydrated.

Electron Shell

The energy level of electrons at a characteristic average distance from the atom's nucleus.

Electron Transport Chain

A group of proteins and enzymes that can receive electrons or hydrogen atoms and transfer them through oxidation-reduction processes. Both aerobic respiration and photosynthesis use electron transport chains. The chains are found in eukaryotic mitochondrial inner membrane systems, as well as bacterial cell membranes and chloroplasts.

The electron is taken from a reduced carbon molecule, which is a TCA cycle intermediate, and oxygen is the final electron acceptor in mitochondria. Dehydrogenases with related flavonoid prosthetic groups, quinones, and cytochromes are all part of the chain. The primary electron chain in chloroplasts connects photosystem I with photosystem II.

Quinones, cytochromes, and other proteins with metal prosthetic groups are all intermediates in the chain. By oxidative phosphorylation or photophosphorylation, much of the energy associated with the transit of electrons along the chain is preserved in the form of ATP.

Electronegativity

A particular atom's attraction to the electrons in a covalent bond.

Electrophoresis

A technique for separating molecules in an electric field based on their differential mobility.

Electrophoretogram

A photographic representation of a column or plate followed by the separation and visualization of a complex mixture through electrophoresis. Typically, the position of the compounds is disclosed by staining with a dye and photographing the result. This is because the gels or colors that form may not be very stable or easy to store for an extended period of time.

Electroporation

Method for delivering recombinant DNA to cells by delivering a short electrical pulse to a solution containing the cells. The pulse generates temporary breaches in the plasma membranes of the cells, allowing DNA to pass through.

Element

Substances which cannot be broken down to any other substance through chemical reactions.

Elimination

The fourth and last stage of food processing in animals where the undigested material is passed out of the body.

Embryo Sac

The female gametophyte of angiosperms which is created when the megaspore grows and divides into a multicellular structure with eight haploid nuclei.

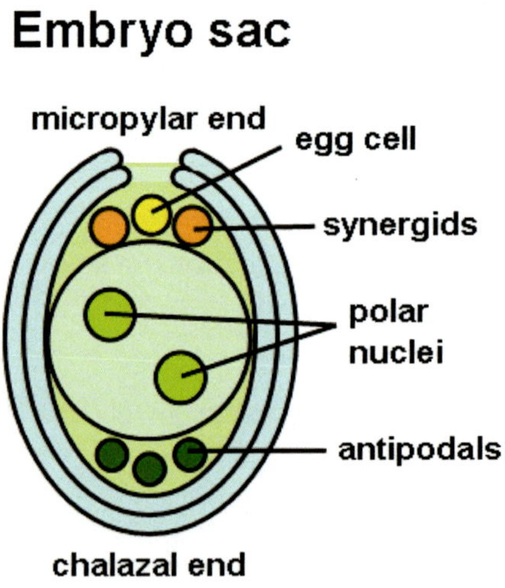

Figure 32. Embryo-Sac.

Source : Image By Flickr

Embryonic Lethal

A phenotypic mutation that causes an embryo or larva to die.

Embryophyte

Alternative word for land plants with specialized reproductive organs having sporophytic embroyo developing in gametophytic tissue of female parent.

Emergent Properties

As the complexity of life increases, the new features which emerge as a result of the arrangement and interactions of parts.

Enantiomer

One of two compounds that are mirror images of each other but due to the presence of an asymmetric carbon they differ in shape.

Endangered Species

A species in danger of extinction throughout all or a significant portion of its range.

Endergonic Reaction

A chemical reaction, where free energy is absorbed from the surroundings and it is nonspontaneous.

Endocrine Gland

Ductless gland by which hormones are secreted directly into the interstitial fluid and further diffused into the bloodstream.

Endocrine System

The internal system of interaction comprising of hormones, the hormone secreting ductless glands, and the molecular receptors on or in target cells that respond to hormones; functions in concert with the nervous system to effect internal regulation and maintain homeostasis.

Endocytosis

The cellular uptake of biological molecules and particulate matter through vesicles' formation from the plasma membrane.

Endoderm

Of the three primary germ layers, the innermost layer in animal embryos; lining the archenteron and giving rise to the liver, pancreas, lungs, and the lining of the digestive tract in species having these structures.

Endodermis

The innermost layer of the cortex that surrounds the vascular cylinder in plant roots.

Endomembrane System

The collection of membranes inside and surrounding a eukaryotic cell, related either by direct physical contact or through the transfer of membranous vesicles; the plasma membrane, the nuclear envelope, the smooth and rough endoplasmic reticulum, the Golgi apparatus, lysosomes, vesicles, and vacuoles are all included in this.

Endometriosis

The condition present due to the presence of endometrial tissue outside of the uterus.

Endometrium

Richly supplied with blood vessels, it is the inner lining of the uterus.

Endophyte

Fungus living inside a leaf or other plant part without harming it.

Endoplasmic Reticulum (ER)

In eukaryotic cells, an extensive membranous network which is continuous with the outer nuclear membrane and composed of ribosome-studded (rough) and ribosome-free (smooth) regions.

Endorphin

Of several hormones produced in the brain and anterior pituitary any hormone that inhibits the pain perception.

Endoskeleton

A hard skeleton which is buried within an animal's soft tissues.

Endosperm

A nutrient-rich tissue formed by the union of a sperm with two polar nuclei during double fertilization, in an angiosperm. It provides nourishment to the developing embryo in angiosperm seeds.

Endospore

A resistant cell with thick-coating, produced by some bacterial cells when they are exposed to harsh conditions.

Endosymbiont Theory

The theory where mitochondria and plastids, including chloroplasts are said to have originated from prokaryotic cells being engulfed by an ancestral eukaryotic cell. The cell which was engulfed and its host cell later evolved into a single organism.

Endosymbiosis

The process in which a unicellular organism (the "host") engulfs another cell, which lives within the host cell and eventually becomes an organelle in the host cell.

Endothelium

The simple squamous layer of cells which line the lumen of blood vessels.

Endothermic

Referring to organisms that are warmed by the generation of heat from their own metabolism. This heat typically maintains a relatively stable body temperature higher than that of the external environment.

Endotoxin

A toxic component of the outer membrane of a particular gram-negative bacteria which is released only when the bacteria die.

Energetic Hypothesis

The idea that the inefficiency of energy transfer along a food chain limits the length of the chain.

Energy

The ability to effect change, particularly via labor (to move matter against an opposing force).

Energy Coupling

The use of energy released from an exergonic event to drive an endergonic reaction in cellular metabolism.

Enhancement

When a system's biological response to two stimuli delivered at the same time is higher than the sum of their separate effects. The term is applied specifically to photosynthesis where the illumination of oxygen producing photoautotroph using a mixture of red and blue light produces oxygen at a faster rate than with either on its own.

Enhancer

A section of eukaryotic DNA that contains numerous regulatory elements and is usually found far from the gene whose transcription it regulates.

Enteric Division

One of three divisions of the autonomic nervous system; that consists of networks of neurons in the digestive tract, pancreas, and gallbladder; generally regulated by the sympathetic and parasympathetic divisions of the autonomic nervous system.

Entrez

An online resource provided by the National Center for Biotechnology Information (NCBI). It organizes GenBank sequences and links them to the literature sources in which they originally appeared.

Enzymatic Hydrolysis

The process in digestion where macromolecules are split from the food through the enzymatic addition of water.

Enzyme

A macromolecule that serves as a catalyst, a chemical agent which increases the rate of a reaction and is not consumed by the reaction. Generally the enzymes are proteins.

Enzyme-Substrate Complex

A temporary complex which is created when an enzyme binds to its substrate molecule(s).

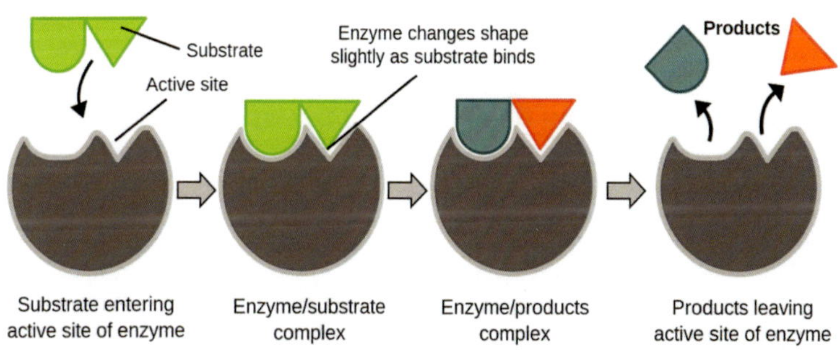

Figure 33. Enzyme-Substrate Binding.

Source : Image By Wikimedia Commons

Epicotyl

The embryonic axis above the point of attachment of the cotyledon(s) and below the first pair of miniature leaves, which is present in an angiosperm embryo.

Epidermis

(1) The dermal tissue system in nonwoody plants that is generally consisting of a single layer of cells that are tightly packed. (2) The outermost layer of cells in an animal.

Epididymis

A tubule which is coiled and located adjacent to the mammalian testis where sperm are stored.

Epigenetic Inheritance

The inheritance of traits which is transmitted through mechanisms that do not involve the nucleotide sequence of a genome directly.

Epinephrine

When secreted as a hormone by the adrenal medulla, the catecholamine that mediates "fight-or-flight" responses to short-term stresses and is also released by some neurons as a neurotransmitter. It is also known as adrenaline.

Epiphyte

A plant which nourishes itself even though it grows on the surface of another plant for support. Generally it grows on the branches or trunks of trees.

Epistasis

When the phenotypic expression of one gene alters that of another independently inherited gene in a type of gene interaction.

Epithelial Tissue

Tightly packed cells in sheets which line the organs and body cavities as well as external surfaces.

Epitope

A tiny, accessible area of an antigen to which an antigen receptor or antibody binds. It is also known as an antigenic determinant.

Equilibrium Potential (Eion)

Using the Nernst equation the magnitude of a cell's membrane voltage which is computed at equilibrium.

Erythrocyte

Also known as the red blood cell, it is a type of blood cell that includes haemoglobin, which transports oxygen.

Erythropoietin (EPO)

A hormone that increases erythrocyte synthesis. When body tissues do not acquire enough oxygen, it is released by the kidney.

Esophagus

A muscular tube that transports food, by peristalsis, from the pharynx to the stomach.

Essential Amino Acid

An amino acid which cannot be produced by an animal itself and hence, must be received in premade form from food.

Essential Nutrient

A substance that an organism can't synthesise from any other source and must thus consume in a preassembled form.

Estradiol

The principal estrogen in mammals, which encourages the development and maintenance of the female reproductive system and secondary sex characteristics.

Estrogen

Any steroid hormone that promotes the development and maintenance of the female reproductive system as well as secondary sex characteristics, such as estradiol.

Estrous Cycle

Female mammals, with the exception of humans and certain other primates, experience this reproductive cycle in which the nonpregnant endometrium is reabsorbed rather than shed, and sexual response occurs only during the mid-cycle at estrus.

Estuary

A region where a freshwater stream or river get merged with the ocean.

Figure 34. Estuaries.

Source : Image By Wikimedia Commons

Ethylene

A plant hormone which is gaseous and is involved in responses to mechanical stress, programmed cell death, leaf abscission, and fruit ripening.

Euchromatin

Less condensed form of the eukaryotic chromatin that is available for transcription.

Eudicot

The member of a clade which contains a vast majority of flowering plants that have two embryonic seed leaves, or cotyledons.

Euglenoid

Like the Euglena, or its relatives, it is a protist, characterized by an anterior pocket which is the point of emergence for one or two flagella.

Euglenozoan

The member of a diverse clade of flagellated protists consisting of predatory heterotrophs, photosynthetic autotrophs, and pathogenic parasites.

Eukaryotic Cell

A type of cell with membrane-enclosed nucleus and organelles. Organisms with eukaryotic cells (protists, plants, fungi, and animals) are called eukaryotes.

Eumetazoan

The member of a clade of animals which have true tissues. Apart from sponges and a few other groups all animals are eumetazoans.

Eutherian

A placental mammal; a mammal whose young complete their embryonic development within the uterus, connected to the mother by the placenta.

Eutrophic Lake

A lake with the presence of a high rate of biological productivity supported by a high rate of nutrient cycling.

Eutrophication

The process through which nutrients, especially phosphorus and nitrogen, become highly concentrated in a body of water, resulting in an increased growth of organisms such as algae or cyanobacteria.

Evaporative Cooling

The process wherein an object's surface begins to cool down during evaporation, a result of the molecules with the greatest kinetic energy changing their state of matter i.e. from liquid to gaseous state.

Evapotranspiration

The complete evaporation of water from an ecosystem, including water transpired by plants and evaporated from a landscape, typically measured in millimeters and estimated for a year.

Evo-Devo

Evolutionary developmental biology is a field of biology that compares developmental processes of different multicellular organisms in order to comprehend how these processes have evolved and how changes can modify existing organismal features or lead to novel features.

Evolution

Descent with modification; the concept of living species being descendants of ancestral species that were contrasting from the present day ones; more narrowly defined as the change in the genetic composition of a population from generation to generation.

Evolutionary Tree

A diagram that reflects a hypothesis about evolutionary relationships among groups of organisms in the form of branches.

Excavata

It is one of the five supergroups of eukaryotes proposed in a present hypothesis of the evolutionary history of eukaryotes. They have unique cytoskeletal features, and some species have an "excavated" feeding groove on one side of the cell body.

Excitatory Postsynaptic Potential (EPSP)

Depolarization or an electrical change in the membrane of a postsynaptic cell which is caused by the binding of an excitatory neurotransmitter from a presynaptic cell to a postsynaptic receptor; making it more likely for a postsynaptic cell to generate an action potential.

Exergonic Reaction

A spontaneous chemical reaction, where there is a net release of free energy.

Exocytosis

The biological molecules' cellular secretion by the fusion of vesicles that contain plasma membrane.

Exon

The portion of DNA from which this sequence was transcribed; also refers to the segment inside a primary transcript that remains in the RNA after RNA processing.

Exoskeleton

A tough encasement on an animal's surface, such as a mollusk's shell or an arthropod's cuticle, that provides protection and attachment places for muscles.

Exotoxin

Even if the pathogen is no longer present, this harmful protein released by a prokaryote or other pathogen causes specific symptoms.

Expansin

Plant enzyme that loosens the fabric of the cell wall by breaking crosslinks (hydrogen bonds) between cellulose microfibrils and other cell wall elements.

ExPASy

A server maintained by the Swiss Institute of Bioinformatics. It is the home of SWISS-PROT, the annotated protein database.

Exponential Population Growth

Represented by a J-shaped curve when population size is plotted over time, it is the growth of a population in an ideal, unlimited environment.

Expression Vector

A cloning vector containing a highly active bacterial promoter simply upstream of a restriction site where a eukaryotic gene can be inserted, allowing the gene to be expressed in a bacterial cell. They're also available that have been genetically modified for utilisation in specific eukaryotic cell types.

External Fertilization

When gametes that have been discharged by parents into the environment fuse.

Extinction Vortex

A downward population spiral where inbreeding and genetic drift combine to cause a small population to shrink and become extinct if the spiral is not reversed.

Extracellular Digestion

Breakdown of food in compartments that is continuous with the external of an animal's body.

Extracellular Matrix (ECM)

The meshwork of glycoproteins, polysaccharides, and proteoglycans generated and secreted by animal cells that surrounds them.

Extraembryonic Membrane

In reptiles and mammals, one of four membranes (yolk sac, amnion, chorion, and allantois) positioned outside the embryo that support the developing embryo.

Extreme Halophile

An organism that thrives in an extremely saline environment, such as the Dead Sea or the Great Salt Lake.

Extremophile

An organism that can survive in harsh environmental conditions where few other species can. Extreme halophiles ("salt lovers") and extreme thermophiles ("heat lovers") are two types of extremeophiles.

E Value (Expect Value)

When performing a BLAST search, you will obtain an E-value for each sequence that is retrieved. An E-value can be thought of as the probability that two sequences are similar to each other by chance. Therefore E-values are best when they are small (e.g. 1 x 10-12).

F

F Factor

The DNA segment in bacteria, confers the ability to form pili for conjugation and associated functions required for the transfer of DNA from donor to recipient. The F factor may exist as a plasmid or be integrated into the bacterial chromosome.

F1 Generation

The first filial, hybrid (heterozygous) offspring that arises from a parental (P generation) cross.

F2 Generation

The offspring which emerges from interbreeding (or self-pollination) of the hybrid F1 generation.

Facilitated Diffusion

A biological process where the rate of uptake of a material across a membrane is increased by a carrier protein present. Facilitated diffusion cannot raise intracellular concentrations of a substance above those seen in the external media, but fast elimination of the item being taken up creates a positive diffusion gradient within the cell.

Facilitation

A type of interaction in which one species benefits the survival and reproduction of another without forming a symbiotic relationship.

Facultative Anaerobe

An organism that makes ATP through aerobic respiration in the presence of oxygen but in its absence switches to anaerobic respiration or fermentation.

FAD

A riboflavin derivative, Flavin adenine dinucleotide; functions as a prosthetic group in flavoproteins participating in oxidation-reduction reactions.

Family

A grouping of closely related or similar genera and are named after the type genus that is characteristic of the whole family. They are divided into subfamilies and grouped into orders.

Fast Block to Polyspermy

Within 1–3 seconds after a sperm connects to an egg membrane protein, the egg plasma membrane depolarizes this Depolarization lasts roughly a minute and stops more sperm from bonding with the egg during that time.

FASTA format

A sequence in FASTA format begins with a single - line description, followed by lines of sequence data. The description line is distinguished from the sequence data by a greater - than symbol (>) in the first column. It is recommended that all lines of text be shorter than 80 characters in length.

Fate Map

The eventual derivatives of specific cells and organs are shown in a territorial diagram of embryonic development.

Fatty Acid

Long-chained carboxylic acid that have three fatty acids connected to a glycerol molecule form a fat molecule, also known as a triacylglycerol or triglyceride, which varies in length and number and placement of double bonds.

Fatty Acid Oxidation

A part of the process of catabolism of stored food reserves in animals and fatty seeds. The first process is hydrolytic breakdown of the three ester linkages to release glycerol and fatty acids, as these reserves are mostly triglycerides. The phosphorylation of glycerol is followed by its oxidation to dihydroxyacetone phosphate and entry into the glycolysis pathway.

The process of beta-oxidation in mitochondria and glyoxysomes breaks down long-chain fatty acids two carbons at a time. Because it creates reduced substrates (e.g., NADPH), which are reoxidized via the terminal electron transport chain to produce ATP, and acetyl-CoA, which can be metabolised by the TCA cycle, this generates ATP.

The fatty acid is first converted to its coenzyme A derivative, which necessitates the use of two molecules of ATP as energy. With the reduction of a FAD-containing flavoprotein, this acetyl-CoA derivative is oxidised to an unsaturated form. More ATP can be produced by reoxidizing the flavoprotein via the terminal electron transport chain.

In a process associated to NADP reduction, the fatty acid derivative's double bond is hydrated and oxidised. The last step requires adding a second coenzyme A molecule to make one acetyl-CoA molecule and one fatty acid-CoA molecule, both of which are two carbons shorter than the starting substrate.

The cycle can be repeated until all of the fatty acids have been converted to acetyl-CoA. The gross yield from one molecule of palmitic acid is 131 ATP molecules, but the net value is 129 since two molecules of ATP are used in the activation of palmitic acid. This shows the value of fats as an energy reserve in biological systems.

Fed Batch System

An anaerobic digester or fermenter that operates in batch mode, with substrate delivered multiple times throughout the run in either solid or concentrated liquid form.

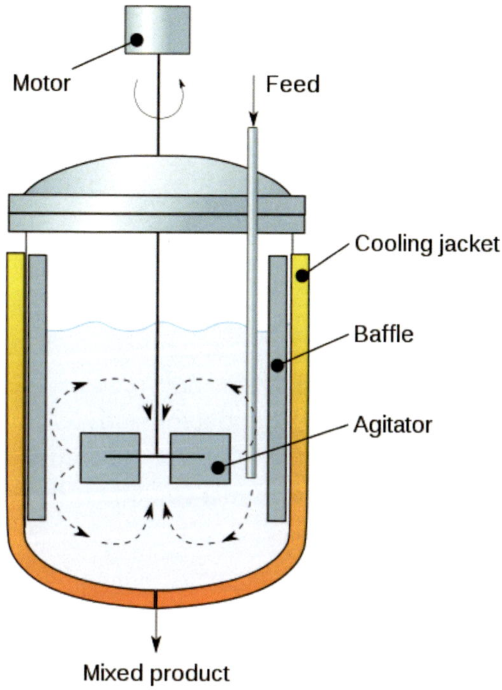

Figure 35. Agitated Vessel.

Source : Image By Wikimedia Commons

Feedback Inhibition

It is a regulation mechanism in which a metabolite produced by a reaction later down the sequence inhibits the activity of an enzyme associated with an early stage of a multi-step pathway. The regulation mechanism could be due to the product interacting to an allosteric location on the enzyme.

Feedback Repression

When an end product of an enzyme's pathway accumulates or is added to a culture media, the enzyme's biosynthesis is blocked.

Fehling's Reaction

A chemical test for reducing sugars and aldehydes. Fehling's solution is made up of two parts: Fehling's I, a cupric sulphate aqueous solution, and Fehling's II, potassium sodium tartrate and sodium hydroxide. The sample solution is boiled after equal volumes of I and II are added. The cupric salt is reduced to create a brick red cuprous oxide precipitate. The solution remains blue in the absence of a reducing chemical.

Fermentation

Degradation of organic molecules by animals or cells to produce chemical energy as ATP in reactions that do not require molecular oxygen in metabolism. It is the only energy-producing activity in anaerobic cells. (2) Descriptive of experimental or commercial operations in which cells or other microorganisms are cultivated in a container (bioreactor or fermenter) in liquid or solid media.

Originally used to denote exclusively anaerobic cultures, the phrase is now commonly used to encompass any cultures of microbes or cells in fabricated vessels.

Fermenter

A vessel or bioreactor used for fermentation. The most common configuration is the completely stirred tank reactor (CSTR). Fermenters range in size from under a litre for bench-scale or experimental systems to hundreds of litres for commercial systems; the only thing they all have in common is that they're closed tanks. Wood, copper, concrete, polymers, aluminium, mild steel, and stainless steel are among the materials utilised in their construction.

If the fundamental construction is made of a less expensive material, a wax or resin lining is frequently employed to prevent against corrosion. Tanks come in a variety of shapes, including domed, square, rectangular, horizontal cylindrical, vertical cylindrical, and conical, with many different forms devised for different types of fermentation or processes.

A standard general-purpose fermenter consists of a cylindrical glass (laboratory scale) or metal (pilot and commercial scale) cylinder with a height/diameter ratio of 2:1 and either a cone-shaped or sloping bottom to facilitate emptying. The fermenter also features ports for fertiliser addition, product removal, and the insertion of measurement and control probes.

Cleaning and sterilisation systems are embedded into larger systems. Gases may be vented or collected through vents in the tank. Aeration and agitation facilities are available as needed. A combination of baffle plates attached inside the fermenter tank and an intermeshing motor-driven rotor is the most typical way of agitation, however alternate methods may be required when the culture's viscosity is high.

These are based on the recirculation of a liquid or gas stream, or the high-pressure injection of air. All aerobic organisms require air (or oxygen) input. In general, this is sparged into the base of the fermenter. The metabolic processes associated with the growth of organisms generate heat, and heat may be transmitted through the stirring device.

In many cases, it is necessary to control the temperature of the fermenter by cooling. Generally this is done using an outer jacket or internal cooling coil through which cold water is passed. In some systems, the fluid is circulated through an external heat exchanger and returned to the vessel.

Systems that use mesophilic or thermophilic organisms may require heating rather than cooling, again using jackets, coils or external heat exchangers. Large fermenters rather than being housed are set out in the open, and in some climates, this will necessitate external insulation. In large systems, access is aided by the provision of catwalks.

Fermenter Agitation

The technique of mixing or stirring contents of a fermenter in order to prevent settling of the organism or product, to facilitate rapid gas exchange (oxygen uptake in particular), and to establish uniform temperature and nutrient distribution.

The most common means of agitation are the use of an impeller rotating at high speed. Internal baffles may be installed in the fermenter to increase the degree of agitation of the broth by creating turbulence. Injection or recycling of gas or liquid streams are two more means of mixing.

Fermenter Cooling

On a commercial scale, cooling towers or refrigeration to supply cold water that is passed via a jacket or internal coil associated with the fermenter are the most frequent techniques of eliminating heat generated during fermentation. Cooling can be a big issue in hot locations when natural water supplies are at temperatures close to or over the optimum for cell growth.

Because humidity is generally fairly high in the summer, when cooling requirements are greatest, a cooling tower is ineffective. Refrigeration, as an option, is rather costly. A huge internal cooling coil wastes a lot of fermenter volume and reduces oxygen transfer rate and mixing quality, which is especially important with viscous broths. A jacket is limited in area to the dimensions of the fermenter. Thus, recirculation through external heat exchangers may be used.

Fermentograph

An automated device for determining the activity of baker's yeast in a sample of dough by measuring carbon dioxide generation.

Fertilization

The fusion of the nuclear material and cytoplasm of two haploid gametes in order to form the diploid zygote, which is a single cell. It usually provides the stimulus for cell division and differentiation to form an embryo.

Foetus

A developing mammal with all the major features of an adult. The fetal stage lasts from the 9th week of gestation until birth, in humans.

Fiber

A cell type which is lignified and reinforces the xylem of angiosperms and functions in mechanical support; a slender, tapered sclerenchyma cell that usually occurs in bundles.

Fibroblast

A type of cell that secretes the protein ingredients of the extracellular fibers in the loose connective tissue.

Fibronectin

An extracellular glycoprotein that helps in attaching animal cells to the extracellular matrix and is also secreted by the animal cells.

Field

In the context of databases, a well-delimited part of an entry containing information of a precise nature (author, date, etc.). You can normally search databases by explicitly targeting specific fields.

Filament

The stalk part of the stamen, in an angiosperm, that produces pollen and is a reproductive organ of a flower.

Filtrate

A cell-free fluid which is extracted from the body fluid by the excretory system.

Filtration

The extraction of water and small solutes, including metabolic wastes, from the body fluid in the excretory systems.

Fimbria

A short, hairlike appendage of a prokaryotic cell that facilitates adhering to the substrate or to other cells.

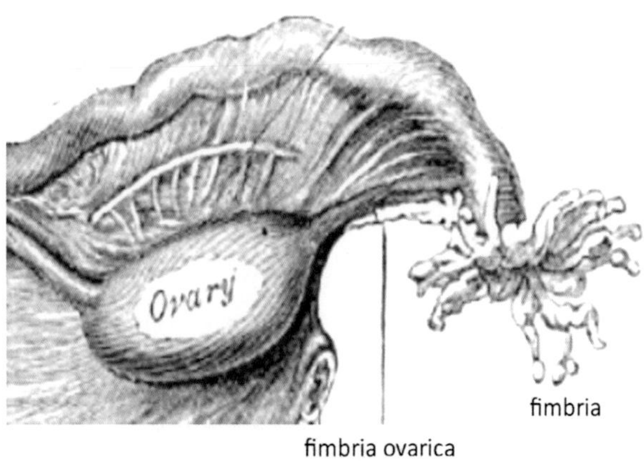

Figure 36. Image Showing Fimbria.

Source : Image By Wikimedia Commons

First Law Of Thermodynamics

The law of conservation of energy: Energy can neither be created nor destroyed, it can only be transferred and transformed.

Fission

The division of an organism into two or more roughly equal-sized individuals.

Fixed Action Pattern

A sequence of unlearned acts that is essentially unchangeable and, once initiated, usually carried to completion and is present in animal behavior,

Flaccid

Limp. Lacking turgor (firmness or stiffness), as in a plant cell in an environment where water tends to escape the cell. (A walled cell becomes flaccid when its water potential exceeds that of its surroundings, resulting in water loss.)

Flagellum

A locomotion-specialized long cellular appendage. Like eukaryotic flagella, motile cilia, have a core ensheathed in an extension of the plasma membrane with nine outside doublet microtubules and two inner single microtubules (the "9 2" arrangement). Flagella in prokaryotes have a unique structure.

Florigen

A flowering signal, most likely a protein, produced in leaves under specific conditions and transmitted to shoot apical meristems, causing them to convert from vegetative to reproductive growth.

Flower

A specialised shoot in an angiosperm with up to four sets of modified leaves bearing sexual reproductive components.

Fluid Feeder

An animal that survives through sucking nutrient-rich fluids from another living organism.

Fluid Mosaic Model

Presently accepted model of cell membrane structure, which envisions the membrane as a mosaic of protein molecules that drift laterally in a fluid bilayer of phospholipids.

Follicle

A microscopic structure present in the ovary containing the developing oocyte and also secretes estrogens.

Follicle-Stimulating Hormone (FSH)

A tropic hormone produced and secreted by the anterior pituitary, that stimulates the production of eggs by the ovaries and sperm by the testes.

Follicular Phase

The segment of the ovarian cycle during which follicles and oocytes are growing and maturing, respectively.

Food Chain

Beginning with producers, the pathway through which food energy is transported from trophic level to trophic level.

Food Vacuole

A sac which is membranous and is formed by phagocytosis of microorganisms or particles that are to be used as food by the cell.

Food Web

In an ecosystem, the interdependent feeding relationships.

Foot

The part of a bryophyte sporophyte that uses transfer cells to collect carbohydrates, amino acids, water, and minerals from the parent gametophyte. (2) One of a mollusk's three primary sections; a muscular structure used for movement.

Forebrain

One of three ancestral and embryonic regions of the vertebrate brain from which thalamus, hypothalamus, and cerebrum are developed.

Fossil

A remnant or impression of an organism which is preserved as it lived in the past.

Founder Effect

The genetic drift that occurs when a few individuals become isolated from a larger population thus forming a new population whose gene pool composition is not reflective of that of the original population.

Fovea

The place on the retina at the eye's center of focus. Here cones are densely concentrated.

FPKM

Fragments per Kilobase of Exon per Million Fragments Mapped; approximates the relative abundance of transcripts in terms of fragments observed from an RNAseq experiment.

Fragmentation

A method of asexual reproduction wherein a single parent divides into parts that regenerate into whole new individuals.

Frameshift Mutation

A type of mutation that occurs when nucleotides are inserted in or deleted from a gene and the number inserted or deleted is not a multiple of three, resulting in the improper grouping of the subsequent nucleotides into codons.

Free Energy

When temperature and pressure are uniform throughout a biological system, the portion of its energy that can accomplish work. The change in free energy of a system (ΔG) is calculated by the equation $\Delta G\ \Delta H - T\Delta S$, where ΔH is the change in enthalpy (in biological systems, equivalent to total energy), T is the absolute temperature, and ΔS is the change in entropy.

Frequency-Dependent Selection

A form of selection in which the fitness of a phenotype is dependent on how common the phenotype is in a population.

Fruit

The mature ovary of a flower that has the function of protecting dormant seeds and often assisting in their dispersal.

Functional Group

A specific configuration of atoms mostly attached to the carbon skeletons of organic molecules that are involved in chemical reactions.

G

G Protein

A GTP-binding protein that transfers signals from a G protein-coupled receptor on the plasma membrane to other signal transduction proteins inside the cell.

G Protein-Coupled Receptor (GPCR)

A signal receptor protein present in the plasma membrane that responds to the binding of a signaling molecule through activation of G protein. It is also called as the G protein-linked receptor.

G0 Phase

The state of non-division which is occupied by cells that have left the cell cycle, sometimes reversibly.

G1 Phase

The first gap, or growth phase, of the cell cycle, that consists of the portion of interphase before DNA synthesis begins.

G2 Phase

The cell cycle's second gap, or growth phase, which is comprised of the interphase period following DNA synthesis.

GA II

Illumina sequencing platform that generates 40 Mb sequence per run with paired read lengths of 70 bp.

Galactose

A sugar that is aldohexose in nature and a constituent of the disaccharide lactose present in milk, as well as occurring in glycolipids, glycoproteins and plant cell wall components. D-Galactose is an epimer of D-glucose.

Galactosidase

An enzyme whose function is conversion of lactose (milk sugar) to glucose and galactose, and also uses other galactose containing compounds as substrates. The enzyme exists in various forms that are highly specific for the galactose residue, but much less specific for the aglycone. The enzyme also acts as a transferase in the formation of oligosaccharides.

Some strains of *E. coli* produce large amounts of this enzyme. Commercial preparations are generally obtained from lactose fermenting yeasts such as Saccharomyces fragilis or fungi such as *Aspergillus niger*, grown on a lactose medium or on whey.

Galactosyl Diglycerides

Lipids found in the chloroplasts of plants. There are mono- and di-galactosyl diglycerides. They are virtually entirel esterified with the di- and tri-unsaturated fatty acids linoleic and a-linolenic acids, with the latter generally predominant. The galactosidic linkage with glycerol is β whereas that between the two galactose residues is α.

Gal (Galactose) Operon

A unit of transcription that contains genes for the three enzymes that convert galactose into glucose-l-phosphate and uridine diphosphate glucose. It is regulated in a manner similar, but not identical to the lac operon.

Gallbladder

The organ meant for the storage of bile and also releasing of the same into the small intestine when required.

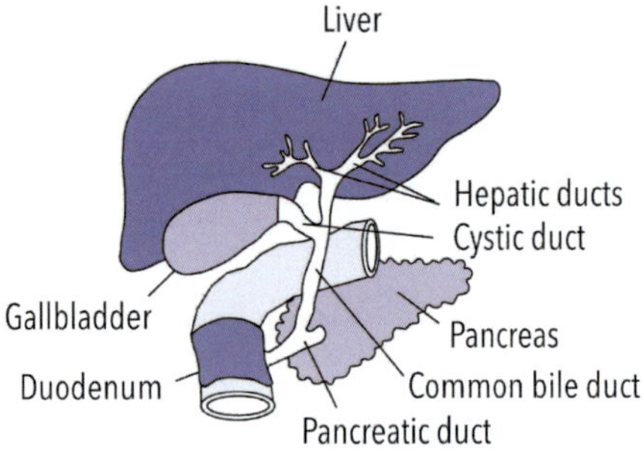

Figure 37. Gall-Bladder.

Source : Image By Flickr

Gametangium

Multicellular plant structure where gametes are formed. Archegonia are the female gametangia, and male gametangia are called antheridia.

Gamete

During sexual reproduction, this haploid cell combines with another gamete to generate a zygote from which the following generation develops In general, such fusion happens between male and female gametes (spermatozoa) (ova). Male gametes are generally motile while female ones are frequently sessile.

Gametophore

A moss gametophyte's mature gamete-producing structure.

Gametophyte

This is one of the stage of life cycle of living organism representing sexual stage. The gametophyte age, which substitutes with a sporophyte age, is the predominant structure in mosses like the bryophytes, however, it becomes diminished in size and life span in further developed vascular plants, the sporophyte becoming prevailing.

In pteridophytes, it is reduced to a little prothallus, though in angiosperms, the gametophyte is addressed by the pollen grain and undeveloped embryo sac.

Gamma-Aminobutyric Acid (GABA)

An amino acid that helps as a CNS synapse in the focal sensory system of vertebrates.

Gamma-Globulin

A kind of serum protein that is recognized based on its electrophoretic mobility and that incorporates both serum immunoglobulins and different proteins without neutraliser movement.

Ganglia

collection of neuronal bodies in peripheral nervous system.

Ganglion Cell

Ganglion cells are cells found in Ganglion transport information from retinal neurons to rest of the brain.

Gap Junction

A type of intercellular channels in animal cells, connect adjacent cells and allows transfer of ions and metabolites between the cells.

Gap penalty

The penalty applied to a similarity score for the introduction of an insertion or deletion gap, the extension of a gap, or both. Gap penalties are usually subtracted from a cumulative score being determined for a comparison of two or more sequences via an optimization algorithm that attempts to maximize that score.

Gas Chromatography

A scientific method used in the study of separation of unknown compounds from sample mixture. The chromatographic interaction relies upon the overall velocities at which the different types go through a long, restricted tube loaded with inactive material, which is covered with a non-unpredictable fluid.

This rate thusly relies upon the segment coefficients of the different parts. Low sub-atomic weight mixtures and gases may likewise be isolated on sections loaded with permeable materials. By warming the segment in a broiler in a customised way, mixtures of fluctuating unpredictability are disintegrated in the arrangement. The interaction is otherwise called gas/ fluid chromatography (GLC).

Figure 38. Gas Chromatography Laboratory.

Source : Image By Wikimedia Commons

Gas Exchange

The take-up of sub-atomic oxygen from the climate and the release of carbon dioxide to the climate.

Gas Holder

A gadget for the capacity of (bio)gas. This might be an extending adaptable sack, a layer over a generally open digester or a drifting gas holder. A drifting gas holder is a chamber shut at its upper end and fixed at the lower end by a fluid, for the most part, water, contained in a tank.

The gas is put away at low tension inside the chamber, over the water level with the chamber rising and falling to oblige the fluctuating gas volume.

Gastric Juice

It is a combination of hydrochloric acid, proteolytic catalysts and bodily fluid released by gastric organs.

Gastrin

A polypeptide chemical, released by the mucosa of the stomach, stimulate the production of hydrochloric acid.

Gastrovascular Cavity

A central cavity with a single opening in the body of specific animals, including cnidarians and flatworms, that helps in both the assimilation and dissemination of supplements.

Gastrula

An undeveloped stage in animal advancement incorporates the development of three layers: ectoderm, mesoderm, and endoderm.

Gastrulation

In animal development, a progression of cell and tissue developments in which the blastula-stage undeveloped organism overlays internal, delivering a three-layered undeveloped organism, the gastrula.

Gated Channel

A transmembrane protein channel that opens or shut considering a specific improvement.

Gated Ion Channel

A gated channel for a particular particle. The opening or shutting of such channels might change a cell's layer potential.

Geiger-Muller Tube (Counter)

A gadget for estimating the movement of a radioactive source or test.

Gel

An actual combination of solid and liquid consisting an inactive polymer structure immersed with a liquid.

Gelatin

A substance acquired by heating up the bones, skins, tendons, and so forth of animals in the water. Comparative substances are additionally acquired from plant deposits. The substance is required to produce gels and is utilised for the culture of certain microorganisms and in plant tissue culture experiment.

Gel Electrophoresis

A scientific method in which charged compounds are isolated and affected by an electrical field (electrophoresis) on a gel. Precast gels of starch, agar, agarose or polyacrylamide areused for separation of compounds.

Separation relies upon the mass/chargeratio of the mixtures and their association with the permeable structure of the gel.

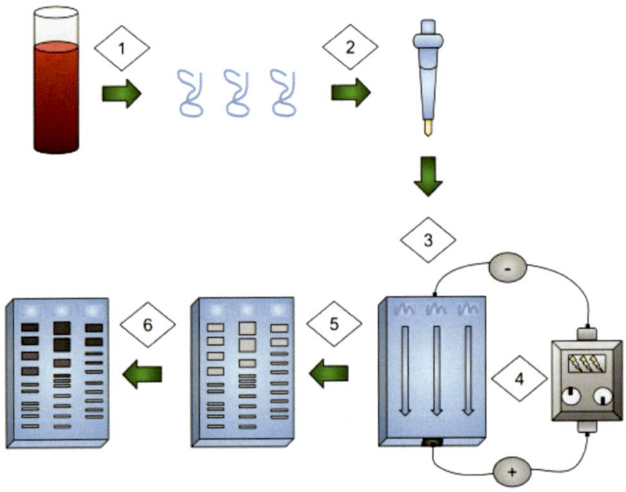

Figure 39. Image Showing Process of Gel Electrophoresis.

Source : Image By Wikimedia Commons

Gel Filtration

A type of chromatography used for separation of molecules based on sub-atomic weight.

Gemmule

These are internal buds found in sponges and certain other mosses plays role in asexual reproduction.

Gene

A unit of inheritance present on chromosome as discrete unit encode for traits. Gene is particulate in nature. Gene concept was postulated by Mendel in his published paper entitled "Experiments in Plant Hybridization" in 1866. The alternate form of gene is known as 'Allele'. Gene pass on from one generation to another as discrete unit. Alleles segregate in process of recombination in Meiosis I during gamete formation. Each gamete receives only one allele from parent. One allele is dominant over the alternative allele of same gene.

Gene Amplification

(1) Multiplication of DNA or nucleotide sequence *in vivo* via Replication or *in vitro* via polymerase chain reaction (PCR).

Gene Annotation

Study of genomic sequence to distinguish protein-coding genes and infer their functions.

Gene Bank

An assortment of cell societies, seeds, frozen sperm or ova, and so forth are kept for the purpose of holding delegated genomes of an organismorganisms.

Gene chips

Also known as gene arrays. Oligonucleotides or cDNA covalently attached directly onto a small glass or silicon chip in organized arrays. Over 50,000 different DNA fragments can be presented on a single chip, providing a high - throughput parallel method of probing gene expression, genotype, or gene function.

Gene Cloning

The development of different copy of a gene. The procedure by which different copy of a plasmid or other cloning vehicle are delivered by embedding the plasmid into an appropriate host equipped for creating numerous copy in mass culture.

The bacterium *E. coli* is frequently utilised as the host organism for gene cloning.

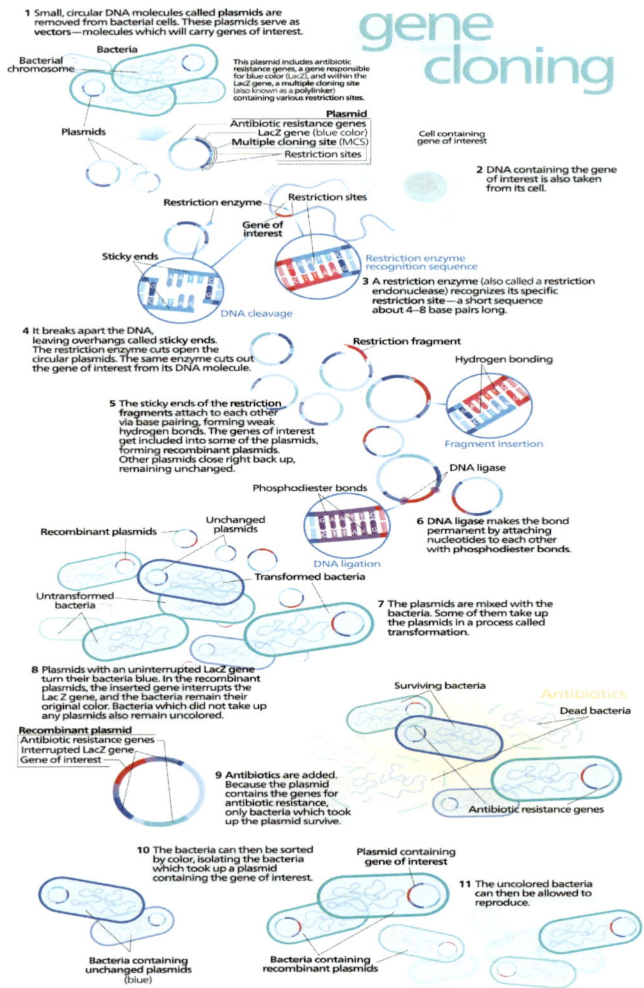

Figure 40. Gene Cloning Process with Bacteria and Plasmids.

Source : Image By Wikimedia Commons

Gene Dosage

The number of duplicates of a particular gene present in the genome of a eukaryotic cell or organism. In diploid cells, the gene is present in two copies on homologous chromosome.

Gene Expression

The process of an simple, complete and fuctional polypeptide or protein from a suitable gene. Development relies upon precise information and translation as well as post-translational handling and compartmentalisation of the initiating polypeptide.

Gene Flow

The transfer of alleles starts with one populace and then onto the next either due to migration of individual to new place and reproduction with the new population or due to movement of gametes.

Gene Isolation

Gene isolation is a method of purifying the DNA sequence using gene cloning or PCR techniques organism.

Gene Library

An arbitrary collection of cloned DNA sequence in several self replicating vectors that represents the individual's entire geneome.

Gene Manipulation

The development of new blends of hereditable material by the inclusion of nucleic acid fragments delivered externally to the cell into any infection, bacterial plasmid or another vector in order to permit their consolidation into a host organism in which they don't normally happen, however in which they are equipped for proceeded with proliferation.

Gene Pool

The total of all duplicates of each type of allele at all loci in each person in a populace. The term is additionally utilised in a more confined sense as the total of alleles for only one or a couple of loci in a populace.

Generation Time

The typical time between one cell division and the following in a prokaryotic cell or the time between the prophase of one cell division and that of the following in a populace of eukaryotic cells. Age times fluctuate from around 20 minutes for quickly developing bacterial population to more than two days for some, plant population.

Generative Nucleus

Generative nuclei are the male gametes; one of the two male nuclei unite with the female gamete in the embrysac to form a zygote and the other male nucleus takes part in double fertilization and forms triploid endosperm.

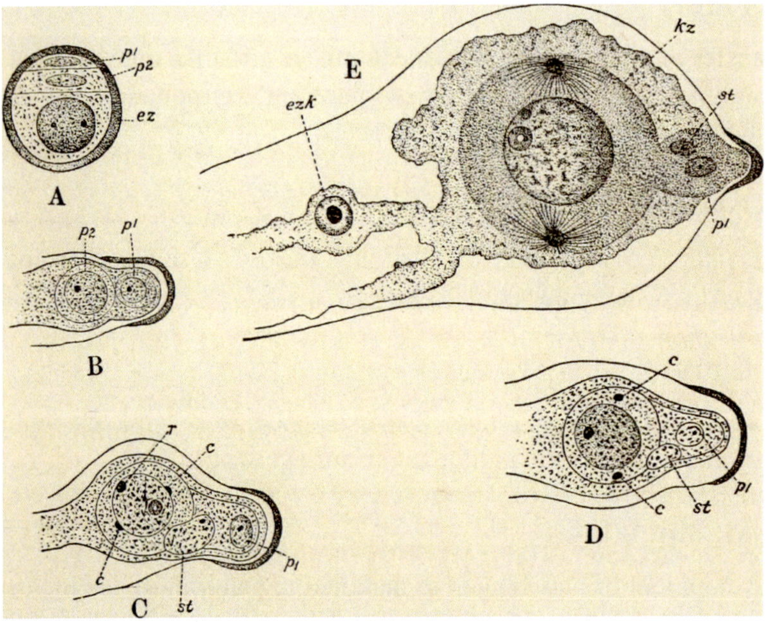

Figure 41. Image Showing Generative Nucleus.

Source : Image By Flickr

Gene Sequencing

A procedure is used to obtain the sequence of bases in the length of DNA coding for a specific polypeptide.

Gene Synthesis

The process of syntheising gene artificially utilizing oligonucleotides through different molecular methods.

Gene Therapy

A procedure embraced to deal hereditarily associated problems or infection.

Genetic Deficiency

The aftereffect of a chromosomal abnormality wherein a part of a chromosome is lost. Deletion are much of the time deadly in homozygous condition.

In individuals, a lack of heterozygous in the short arm of chromosome 5 is related to microencephaly, extreme development irregularities and mental impediments. The deletion of part of chromosome 22 is related to persistent myeloid leukaemia.

Heterozygous lack in human chromosomes 4, 13 and 18 are related to extreme physical and mental debilitations.

Gene-For-Gene hypothesis

This hypothesis is proposed by Flor in 1956 on his studies on flax rust of Linseed crop. For every resistance gene in host plant there is matching virulent gene in pathogen.

Genetic Drift

Genetic drift is process when allelic frequency of any population changes over time randomly.

Genetic Engineering (in Recombinant DNA Technology)

The procedures used to isolate, splice, and manipulate DNA outside the cell. Genetic engineering allows a recombinantly engineered DNA segment to be introduced into a foreign cell or organism and to be able to replicate and function normally.

Genetic Map

Genetic Map is prepare to locate the genetic marker and determine their sequence using genetic linkage or recombination frequency.

In bigger genomes, not every one of the gene might have been located, and there might be lapse in the hereditary guide. In the genome of SV40, DNA codes for five proteins: VPl, VP2, VP3, t antigen and T antigen. A portion of these proteins contain indistinguishable amino acid sequence and are coded for by a similar stretch of DNA.

The mRNAs determining the five proteins represent practically the whole length of the genome. Interestingly, in length of the eukaryotic chromosomes, there are enormous regions of DNA that don't code for amino acids.

Genetic Marker

A sequence of DNA that are linked with studied traits with little or no chances of recombination. The associated genetic marker with the gene of interest transmits together from one generation to another cells.

Genetic Modification

An interaction that results in a modification of the hereditary make-up of a populace. Strategies for genetic alteration include cloning (intervened by wild-type plasmids) in prokaryotes; *in vivo* modifications of transposable components; *in vitro* DNA recombination procedures; protoplast combination; the utilisation of mutagens; hybridisation by conventional reproducing methods.

In prokaryotes, the utilisation of mutagens is the most generally adopted procedure for the development recombinant strains. In eukaryotes, the most common method is conventional reproduction and hybridisation.

Genetic Profile

An individual's have typical profile of hereditary markers, distinguished most frequently today by PCR or, beforehand, by electrophoresis and nucleic acid tests.

Genetic Recombination

The transfer of genes between two chromosomes. This happens normally in crossing over. *In vitro* hereditary recombination requires the extraction of DNA, or the development of cloned DNA from an appropriate mRNA, from one source and the inclusion in another.

Genetic Variation

Variation present among people due to the difference in DNA sequence of their genome.

Genetically Modified (GM) Organism

An organism with foreign DNA transferred through recombinant DNA technology; otherwise called a transgenic organism.

Genetics

The study of heredity and inheritance of characters in living organism..

Genome

The total hereditary material of an organism or individual; the total supplement of an organism's that constitute both coding and non-coding sequences.

Genome Browser

Genome browser is a graphical interface for display of information from a genomic data of biological organisms.

Genome-Wide Association Study

The comparative study of genomes of several induividuals to track down the genes responsible for the inheritance of specific trait. This involves large scale sequencing data to analyse and interpret the information with respect to specific traits.

Genomic Imprinting

It is epigenetic phenomena wherein expression of an allele in progenies relies upon whether the allele is acquired from the male or female parent.

Genomic Library

A collection of cell clones containing all the DNA fragments from a genome, each inside a plasmid, BAC, or another cloning vector.

Genomics

The investigation of total hereditary material of an organism including the comparative analysis with other organisms' genomes.

Genus

A unit of characterisation. A variety incorporates those firmly related species that share a conspicuous series of attributes for all intents and purposes. In the binomial arrangement of classification of living organism, the principal name given is that of the Genus followed by the species name.

Geographic Variation

Differnece between the genetic material of geographically separate populaces or populace subgroups.

Geologic Record

The division of Earth's set of experiences into time-frames, assembled into three ages — Archaean, Proterozoic, and Phanerozoic — and further partitioned into periods, periods, and ages.

Germ Layer

One of the three fundamental layers in a gastrula that will shape the different tissues and organs of a creature's body.

Gestation

Pregnancy; is the condition of conveying creating youthful inside the female regenerative parcel.

GFF (Generic Feature Format)

A data format for identifying the features of a sequence that involves data featured in a tab-delimited file with one feature per line to enable text manipulation and data analysis tools that work with tabular data.

Gibberellin

Any of a class of related plant chemicals that stimulate development in the stem and leaves, trigger the germination of seeds and breaking of bud torpidity, and (with auxin) induce natural plant improvement.

Glans

The structure at the tip of the clitoris or penis is associated with sexual excitement.

Glia (Glial Cells)

Cells of the sensory system help, control, and increase the elements of neurons.

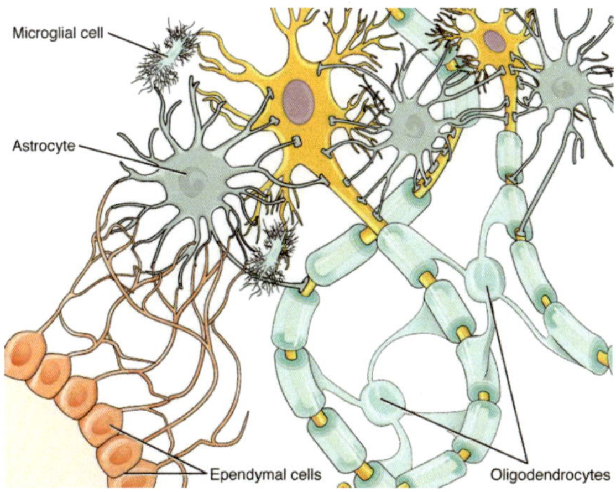

Figure 42. Glial Cells of The CNS.

Source : Image By Wikimedia Commons

Global Climate Change

Expansion in temperature and change in weather conditions from one side of the world to the other, due for the most part to expanding environmental CO2 levels from the consumption of non-renewable energy sources. The expansion in temperature, called a dangerous atmospheric deviation, is a significant part of worldwide environmental change.

Global Ecology

The investigation of the working and appropriation of organisms across the biosphere and what the territorial trade of energy and materials means for them.

Glomeromycete

Individuals from the contagious phylum Glomeromycota, are described by an unmistakable stretching type of mycorrhizae called arbuscular mycorrhizae.

Glomerulus

A bundle of vessels encompassed by Bowman's capsule in the nephron and filling in as the site of filtration in the vertebrate kidney.

Glucagon

A chemical released by pancreatic alpha cells raises blood glucose levels. It advances glycogen breakdown and the arrival of glucose by the liver.

Glucocorticoid

A steroid chemical that is discharged by the adrenal cortex impacts glucose digestion and insusceptible ability.

Glutamate

An aminoacid that helps as a synapse in the central nervous system.

Glyceraldehyde 3-Phosphate (G3P)

A three-carbon starch that is the intermediate metabolite of the Calvin cycle.

Glycogen

A multibranch polysaccharide represents storage form of glucose in the liver and muscle of the animals.

Glycolipid

A lipid with at least one covalently connected starch.

Glycolysis

A progression of responses that eventually parts glucose into pyruvate. Glycolysis happens in practically all living cells, filling in as the beginning stage for maturation or cell respiration.

Glycoprotein

A protein with at least one covalently connected starch.

Glycosidic Linkage

A covalent bond is developed between two monosaccharides by a lack of hydration response.

Golden Glga

A biflagellated, photosynthetic protist named for its tone, which results from its yellow and earthy coloured carotenoids.

Golgi Apparatus

An organelle in eukaryotic cells comprising of heaps of level membranous sacs that change, store, and course results of the endoplasmic reticulum and orchestrate a few items, outstandingly noncellulose starches.

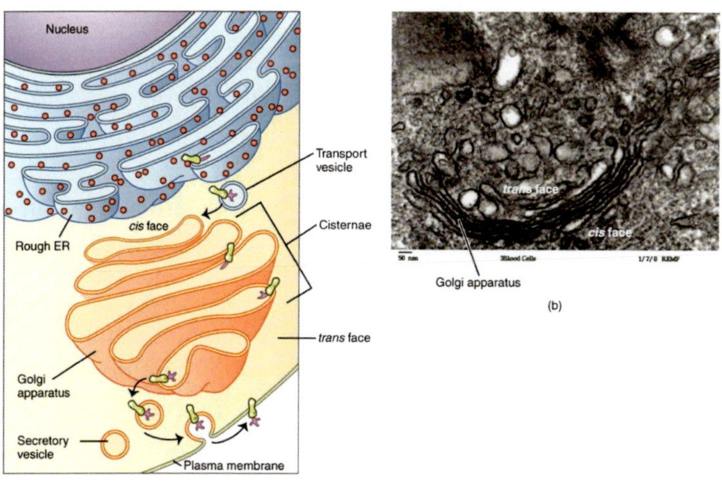

Figure 43. (A) The Golgi Apparatus Manipulates Products from The Rough ER, And Also Produces New Organelles Called Lysosomes, (B) An Electron Micrograph of The Golgi Apparatus.

Source : Image By Wikimedia Commons

Gonads

The male and female sex organs; are the gamete-delivering organs in many animals.

Grade

A gathering of organic entities that share a similar degree of authoritative intricacy or offer a key transformation.

Graded Potential

In a neuron, a change in the layer potential that has a plentifulness corresponding to flag strength rots as it spreads.

Gram Stain

A staining technique that recognizes two various types of bacterial cell walls; might be utilised to assist with deciding clinical reaction to contamination.

Gram-Negative

Portraying the group of microscopic organisms that have a cell wall that is fundamentally more complex and contains less peptidoglycan than the cell mass of gram-positive microorganisms. Gramnegative microscopic organisms are mostly more poisonous than gram-positive microorganisms.

Gram-Positive

Portraying the group of microscopic organisms that have a cell wall that is fundamentally less complex and contains more peptidoglycan than the cell mass of gram-negative microorganisms. Gram-positive microscopic organisms are typically less poisonous than gram-negative microorganisms.

Granum

A pile of layers limited thylakoids in the chloroplast. Grana work in the light responses of photosynthesis.

Grey Matter

Area of dendrites and collection of neuron cell bodies inside the CNS.

Green Alga

A photosynthetic protist, named for green chloroplasts that are comparative in construction and shade organisation to those of land plants. Green growth is a paraphyletic group, a part of whose individuals are more strongly connected with land plants than they are to other green alga.

Greenhouse Effect

The warming of Earth is because of the environmental accumulation of carbon dioxide and certain different gases, which assimilate reflected infrared radiation and reradiate some of it back toward Earth.

Gross Primary Production (GPP)

The absolute essential creation of a biological system. The total amount of organic carbon produce by plants in ecosystem in given time duration.

Ground Tissue System

Plant tissues that are neither vascular nor dermal, satisfy an mixture of role, like photosynthesis, and support.

Growth Factor

(1) A protein that should be available in the extracellular climate (culture medium or animal body) for the development and ordinary improvement of particular kinds of cells. (2) A neighbourhood controller that follows up on adjacent cells to stimulate cell multiplication and separation.

Growth Hormone (GH)

A chemical that is delivered and discharged by the foremost pituitary and that has both direct (nootropic) and indirect role for a wide array of issues.

Guard Cells

The two cells flank the stomatal pore and control the opening and closing of the pore.

Gymnosperm

An avascular plant that bears naked seeds — seeds not enclosed in protective coat.

H

Hair Cell

A mechanosensory cell that modifies results in the sensory system when hairlike projections on the cell surface are dislodged.

Half-Life

How much time does it take for a radioactive isotope to decay

Hamilton's Rule

The rule is that for regular determination to incline toward a philanthropic demonstration, the advantage to the beneficiary, depreciated by the coefficient of relatedness, should surpass the expense for the altruist.

Haploid Cell

A cell containing just a single set of chromosomes (n).

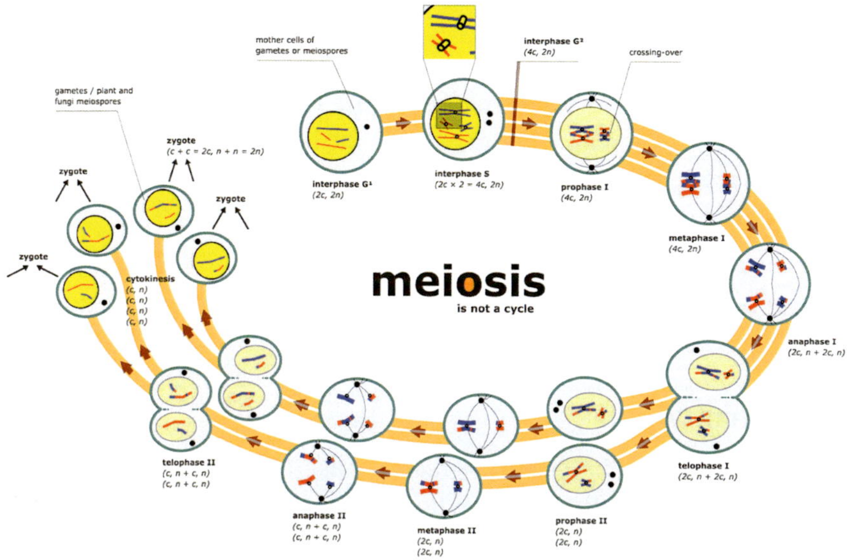

Figure 44. Meiosis Is a Process in Which A Diploid Cell Divides Itself Into 4 Haploid Cells.

Source: Image by Wikimedia Commons

Haplotype

A collection of alleles in one individual those are located on one chromosome. Alleles within a haplotype are often inherited as a single unit from one generation to the next. In SNP studies, haplotypes refer to a group of genomic variations found repeatedly in many individuals within a population.

Hardy-Weinberg Principle

The rule that frequencies of genes and genotypes in a populace stay consistent overtime provided there is no selection, mutation, migration and genetic drift.

Haustorium

In specific advantageous parasites, a particular hypha can infiltrate the tissues of host organism.

Heart

A strong pump that uses metabolic energy to hoist the hydrostatic tension of the circulatory blood (blood or hemolymph). The blood then, at that point, streams down a tension inclination through the body and in the long run gets back to the heart.

Heart Attack

The harm or passing of cardiovascular muscle tissue comes about because of drawn out blockage of at least one coronary supply route.

Heart Murmur

A murmuring sound that most frequently results from blood spurting in reverse through a defective valve in the heart.

Heart Rate

The recurrence of heart constriction (in pulsates each moment).

Heat map

Graphical representation of data where the individual values contained in a matrix are represented as colors.

Heat-Shock Protein

A protein that protects different proteins during heat pressure. Heat-shock proteins are found in plants, creatures, and microorganisms.

Helicase

An enzyme that untwists the twofold helix of DNA at replication forks, isolating the two strands and making them accessible as template strands.

Helper T Cell

A type of T cell that, when initiated, secretes cytokines that advance the reaction of B cells (humoral reaction) and cytotoxic T cells (cell-interceded reaction) to antigens.

Haemoglobin

An iron-containing protein in red platelets that reversibly ties oxygen.

Hemolymph

In spineless animals with an open circulatory structure, the body is liquid that washes tissues.

Haemophilia

A human hereditary disorder caused due to X-linked recessive allele leading to reduced blood-coagulating protein; described by excessive bleeding following injury.

Hepatic Portal Vein

An enormous vein that transport blood from the small digestive system (intestines, gallbladder, pancreas and spleen) to the liver..

Herbivore

Animals that eats only plants for all their life.

Herbivory

It is consumption of plant or alga by animals.

Heredity

The transmission of characteristics from one generation to next.

Hermaphrodite

An individual organism that has both male and female reproductive organs.

Hermaphroditism

A condition wherein an individual has both female and male reproductive organs.

Heterochromatin

It is a region of eukaryotic chromatin that remains densely packed found mostly in the centromere or telomeric region of chromatin. It contains mostly repetitive DNA sequences. The Heterochromatin region is highly conserved.

organismHeterocyst

Differentiated cell of cyanobacteria that takes part in nitrogen fixation.

Heterokaryon

A fungal mycelium that contains at least two haploid nuclei in their cell.

Heteromorphic

Referring to a condition in the existing pattern of plants and certain green algae in which the sporophyte and gametophyte phases are present in their lifecycle.

Heterosporous

Referring to a plant-animal category that has two sorts of spores: microspores, which develop into male gametophytes, and megaspores, which develop into female gametophytes.

Heterotroph

An organismorganism doesn't prepare own food and depends on other organisms for their organic carbon supply.

Heterozygote Advantage

It is a condition where heterozygous genotype of an organism has high relative fitness than the homozygous genotype (Dominant/ recessive).

Heterozygous

Having two alternate alleles for a given gene.

Hibernation

It is an inactive physiological state present in some organisms wherein metabolic depression occurs for a specific time period to conserve energy for surviving adverse weather condition..

High-Density Lipoprotein (HDL)

A molecule in the blood is comprised of thousands of cholesterol particles and different lipids bound to a protein. HDL absorbs cholesterol and transport it to the liver.

High-throughput

Methods that produce large volumes of data and can process many samples quickly. Robots and computerized data collection are common in high-throughput methods.

Hindbrain

One of three genealogical and early stage districts of the vertebrate cerebrum; forms into the medulla oblongata, pons, and cerebellum.

HiSeq

Illumina sequencing platform that generates 120 Gb sequence per run with paired read lengths of 125 bp.

HiSeq2500

Illumina sequencing platform that generates 1,000 Gb sequence per run with paired read lengths of 125 bp.

Histamine

Organic nitrogenous molecule involved in immune reaction and also takes part in various physiological responses..

Histone

A protein with a high extent of emphatically charged amino acids that ties to the adversely charged DNA and assumes a key part in chromatin structure.

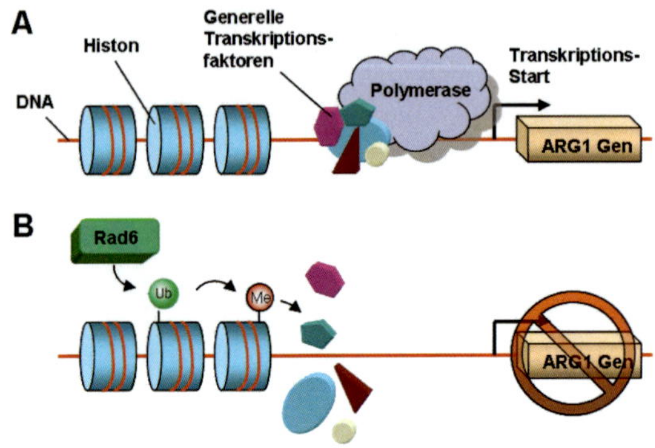

Figure 46. Rad6-Dependent Histone Ubiquitination, Subsequent Histone Methylation and Silencing Of The Arg1 Gene In Yeast.

Source : Image By the Wikimedia Commons

Histone Acetylation

The connection of acetyl aggregates to specific amino acids of histone proteins.

HIV (Human Immunodeficiency Infection)

The human immunodeficiency virus that causes Acquired immunodeficiency syndrome (AIDS), a chronic disease affects immune system severely. HIV is a retrovirus.

HMM (Hidden Markov Model)

A mathematical formulation of a succession of hidden, mutually exclusive properties (such as intron or exon) associated with one sequence or a multiple sequence alignment. In the context of proteins HMM is used interchangeably with the words 'motifs' or 'profiles'. In the context of DNA, HMMs are used to predict the location and structure of genes.

Holoblastic

Alluding to a sort of cleavage wherein there is a finished division of the egg; which happens in eggs that have little yolk (like those of the ocean imp) or a moderate amount of yolk (similar to the frog).

Homeobox

A 180-bp DNA coding for 60 amino acid DNA binding protein that are generally conserved in organisms. Related homeo genes present in plants and yeasts.

Homeotic Gene

The genes that control the arrangement and spatial association of anatomical structure in different organisms, plants, insects and mammals.

Hominin

An individual from the tribe of Hominini (family Hominidae, order Primates). Hominins incorporate Homo sapiens and our predecessors, a group of extinct animal categories that are more firmly connected with us than chimpanzees.

Homologous Chromosomes

A couple of chromosomes of a similar length, centromere position, and staining design have qualities for similar characteristics at comparing loci. One homologous chromosome is acquired from the organismindividual's father, the other from the mother. Additionally, called homologs, or a homologous pair.

Homologous Structures

Structures in various species that are comparative as a result of the normal family line

Homology

Comparability in attributes comes about because of a common family line.

Homoplasy

A comparative (practically equivalent to) structure or sub-atomic arrangement that has developed autonomously in two species.

Homosporous

Alluding to a plant-animal category that has a solitary sort of spore, which ordinarily forms into a sexually unbiased gametophyte.

Horizontal Cell

A neuron of the retina modulates the information flow between photoreceptor cells and bipolar cells.

Horizontal Gene Transfer

The exchange of genes starts with one genome and then onto the next through components, for example, transposable element, plasmid, viral movement, and maybe combinations of various gene molecules.

Hormone

In multicellular organisms, one of many kinds of secreted substances that are chemical messangers, pass into blood stream, and flow into different parts of the body, to exerts its biological funtion.

Hornwort

A little, herbaceous, nonvascular plant that is an individual from the phylum Anthocerophyta.

Figure 47. Mystery Hornwort.

Source : Image By Flickr

Host

is and organism that provide a congenial environment for other organisms to grow.

Host Range

The predetermined number of species whose cells can be contaminated by a specific infection.

Human Chorionic Gonadotropin (HCG)

A chemical secreted by the syncytiotrophoblastic cells of the placenta that stimulates the corpus luteum of the ovary to produce progesterone during the initial three months of pregnancy.

Human Genome Project

A worldwide cooperative work to to MAP and sequence the DNA of the whole human genome.

Humoral Immune Response

The part of versatile insusceptibility includes the initiation of B cells and that prompts the development of antibodies, which safeguard against microscopic organisms and infections in body liquids.

Humus

Disintegrating natural material that is a part of dirt. Huntington's infection A human hereditary sickness brought about by a predominant allele; described by wild body developments and degeneration of the sensory system; typically, lethal 10 to 20 years after the beginning of side effects.

Huntington's Disease

A human hereditary infection brought about by a predominant allele; described by wild body developments and degeneration of the sensory system; typically, lethal 10 to 20 years after the beginning of side effects.

Hybrid

Posterity outcomes from the mating of people from two distinct species or from two genuine reproducing assortments of similar species.

Hybrid Zone

A geographic district wherein individuals from various species meet and mate, delivering some posterity of blended family line in any event.

Hybridization

In hereditary context, the mating, or crossing, of two pure homozygous reproducing genotypes.

Hydration Shell

The circle of water particles around a disintegrated particle.

Hydrocarbon

A natural particle comprising just carbon and hydrogen.

Hydrogen Bond

It is the electrostatic force of attraction hydrogen atom which is covalend bond between hydrogen bond donor (electronegative atom) and hydrogen bond acceptor (electronegative atom with lone electron)

Hydrogen Ion

A solitary proton with a charge of 1+. The separation of a water particle (H2O) prompts the age of a hydroxide particle (OH-) and a hydrogen particle (H+); in water, H+ isn't viewed as alone however connects with a water atom to frame a hydronium particle.

Hydrolysis

A synthetic response that breaks connections between two particles by the expansion of water; capacities in dismantling of polymers to monomers.

Hydronium Ion

A water particle that has an additional proton bound to it; H3O+, is ordinarily addressed as H+.

Hydrophobic

Having no affection for water; tending to combine and shape drops in water.

Hydrophobic Interaction

A sort of electronegative molecule interactioninteraction is caused when particles that don't blend in with water combine to prohibit water.

Hydroponic Culture

A technique wherein plants are sown in mineral solution as opposed to in soil.

Hydroxyl Group

A substance group comprising of an oxygen molecule joined to a hydrogen particle. Particles having this structure are dissolvable in water and are called alcohols.

Hymen

A slender layer that part of the way covers the vaginal opening in the human female. The hymen is cracked by sex or another enthusiastic movement.

Hyperpolarization

An adjustment of a cell's layer is expected to such an extent that within the film turns out to be more negative compared with the outside. Hyperpolarization diminishes the opportunity that a neuron will communicate a nerve motivation.

Hypersensitive Response

A plant's restricted protection reaction to a microorganism, including the killing cells around the site of contamination.

Hypertension

A problem wherein circulatory strain remains strangely high.

Hypertonic

Alluding to an answer that, while encompassing a cell, will make the cell lose water.

Hypha

One of many associated fibres that by and large make up the mycelium of an organism.

Hypocotyl

In an angiosperm undeveloped organism, the early stage pivot underneath the place of connection of the cotyledon(s) or more the radicle.

Hypothalamus

The central piece of the vertebrate forebrain; capacities in keeping up with homeostasis, particularly in organising the endocrine and sensory systems; secretes chemicals of the back pituitary and delivers factors that control the foremost pituitary.

Hypothesis

A testable clarification for a bunch of perceptions considering the accessible information and directed by inductive thinking. Speculation is smaller in scope than a hypothesis.

Hypotonic

Alluding to answer that, while encompassing a cell, will make the cell take up water.

I

Idioblast

A cell that contrasts with the encompassing cells inside a composite tissue.

Ideogram

An outline or picture of the total chromosome compliment of a cell. It is normal for an organismorganism as far as chromosome number and chromosome phenotype.

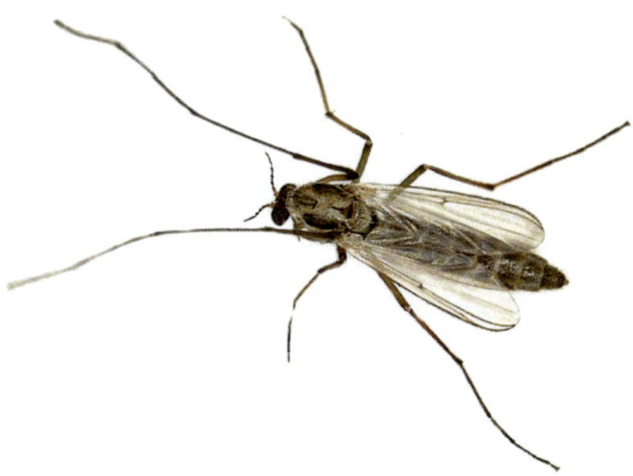

Figure 48. Chironomus Spec. (Chironomidae) - (Imago), Elst (Gld) - De Park, The Netherlands.

Source : Image By Wikimedia Commons

Imbibition

The adsorption of water by colloids, gels and comparative substances. It is an actual interaction as opposed to a metabolic one. Imbibition is normal in plants, and in seeds specifically. As water is assimilated the substance of the cotyledons or endosperm enlarges causing a high imbibition pressure.

This prompts the crack of the seed coat and the arrival of the creating root and additionally plumule. Expansions in the new weight of extending cells in the meristematic district of creating plants are additionally part of the way because of imbibition and part of the way because of assimilation.

Immigration

The convergence of new people into a populace from different regions.

Immobilization

Physical or synthetic interaction is used to fix catalysts, microorganisms and societies of plant or animal cells, or organelles got from these sources, onto a strong help or trap them in a strong framework. Such frameworks are utilized as biocatalysts of inconsistent stream frameworks (e.g., pressed segment reactors or extended bed frameworks).

Immune Modulator

A compound that interferes with the development of a counteracting agent considering an antigen. Insusceptible modulators are utilised to decrease the chance of a host dismissing a heterograft.

Immune Response

The occasions that happen in people and other vertebrate animals when the body is attacked by unknown protein. It is reflected by the development of antibodies and might be invigorated by an irresistible organismorganism or microbes (microorganisms, yeast, growths, protozoa, and so forth), translocated material, immunisation, sperm or even the host's own tissue.

Immune System

An animal's body's arrangement of protections against antigens that cause infection.

Immunity

The capacity to oppose contamination or reject unfamiliar tissue (protein) is presented by the development of explicit antibodies in animals. The dynamic insusceptibility is because of the improvement of an invulnerable reaction following the excitement of the resistant response by an antigen.

It might create openness to contamination or be instigated by immunization. Insusceptibility is portrayed by the development of humoral counteracting agents or lymphocytes containing cell-bound antibodies.

Immunization

A procedure used to instigate insusceptible protection from a particular infection in people and different warm-blooded animals by intentional openness to an antigen to raise antibodies to that antigen. It is accomplished by infusion of a combination of antibodies to the infection (uninvolved insusceptibility) or by immunization which invigorates the invulnerable response to deliver explicit antibodies.

Immunisation presents longer-enduring dynamic insusceptibility, though uninvolved invulnerability gives just impermanent insurance since the remotely determined counteracting agent is processed by the beneficiary.

Immunoaffinity

Immunocytochemical Method

A technique that utilises a sign compound connected to a particular counteracting agent to find a protein in a cell. The cells are responded with the marked counteracting agent and afterwards inspected under the magnifying instrument (after improvement of the variety), analyzed for fluorescence or exposed to autoradiography, as suitable to the naming procedure utilised.

Immunodeficiency

A problem wherein the capacity of an insusceptible framework to safeguard against microorganisms is inadequate or missing.

Immunodiagnostic

An insightful test in light of the exceptionally unambiguous interaction between a counteracting agent and an antigen. Interaction is connected to a technique for evaluating the outcome. Most immunodiagnostics depends on radioimmunoassays. Different techniques incorporate immunodiffusion and immunoelectrophoresis.

Immunoelectrophoresis

A scientific procedure utilised for differentiating antigens isolated from a combination of proteins (e.g., blood serum) by electrophoresis. It is a separation and identification of proteins due to differences in electric charge and reactivity with antibodies.

Immunofluorescence

An insightful procedure wherein a fluorescent colour is complexed with a particular counteracting agent to the protein of interest. The counteracting agent buildings with the protein and the complex shape is perceived utilising a light magnifying instrument or gadgets, for example, computerised cell sorters where the fluorescence is recorded by a photomultiplier.

The procedure is utilised to find a given protein or compound in a tissue segment, to distinguish layer bound proteins, or to recognize or identify the quantity of a particular kind of cells in a blended populace. In blended with PC based counters or picture analysers, exceptionally point by point data can be acquired because of the particularity of the name and the responsiveness of identification gadgets.

Immunoglobulin

Any of the class of proteins that capacity as antibodies. Immunoglobulins are classified into five significant classes that contrast in their dissemination in the body and antigen removal exercises.

Immunological Tolerance

A resilience towards a specific antigen in a singular human or warm-blooded creature. It results from a disappointment of the insusceptible reaction. The resistance might be because of hereditary variables, openness to antigens during fetal life, openness to ingested antigens or a general disappointment of the insusceptible reaction as in specific sicknesses like AIDS.

Immunology

The investigation of the nature and conduct of the insusceptible reaction and invulnerable framework in people and different animals.

Immunomodifier

A synthetic or natural specialist modifies the insusceptible reaction, for example, by invigorating the arrangement of counter acting agent delivering cells or by modifying the phone properties by restricting to external cell layers.

Immunotoxin

A counterfeit item is acquired by intertwining an immune response with a poison of plant or bacterial beginning. Such mixtures are utilised to convey possible pharmacological specialists to explicit regions and to cancer cells specifically.

Impeller

A multibladed gadget, pivoted at high velocity, in a fermenter to help the pace of oxygen move and to keep the cells in suspension.

Figure 49. Impeller For a Three-Stage Turbo-compressor Used to Produce Compressed Air. Built In 1985. MAN Museum, Augsburg.

Source : Image By Wikimedia Commons

Implantation

The interaction by which the prepared ovum of well-evolved animals becomes embedded in the endometrium of the uterus toward the beginning of pregnancy.

Imprinting

An educational experience found in animals, and birds specifically, that outcomes in a particular reaction to a specific article. Frequently the reaction, which happens at an early age, brings about the article being followed.

In nature, this reaction is instigated by the parent. Be that as it may, under counterfeit circumstances an animal can figure out how to follow different animals.

Impulse

A sign that is proliferated through the sensory system along the axons of neurones. It is the consequence of changes in the cell layer properties which cause an expected inclination across the film. The trademark changes potentially related to motivation (activity potential) can be recorded utilizing microelectrodes.

The limit improvement expected to deliver a motivation is instigated by a neighbourhood depolarization, bringing about an end-plate potential, excitatory postsynaptic potential or receptor potential, as indicated by the site where it happens.

When started the change is proliferated along with the layer, first as an internal development of sodium particles, which turns around the potential from - 70 mV to +30 mV, then, at that point, as an outward development of potassium particles, which reestablishes the resting potential.

Following the section on motivation, the layer can't be invigorated again until it recuperates to the resting state. Every one of the driving forces is a similar size, subsequently, data is reliant upon the recurrence and complete number of motivations summoned by a particular improvement.

Inbreeding

The mating of genetically related people. Inbreeding brings about an ever-evolving decline in the level of heterozygosity of the populace which is reflected by a diminishing in the variety found in the aggregates of a given populace. Inbreeding may likewise bring about the appearance of deleterious recessive alleles.

Incipient Species

Populaces that are too unmistakable to be in any way considered as subspecies of similar species, however not adequately separated to be viewed as various species; additionally, called semi species.

Inclusion Body

A construction, comprising of numerous infection particles implanted in a protein body, shaped in plant cells contaminated by caulimoviruses.

Incompatability

The consequence of a mechanical, hereditary or physiological interaction between two organisms of other genders that forestalls or brings about a incompatibility. In plants, incompatibility mechanism exist to forestall self-pollination in hermaphrodite flowers or individual plants that bear both male and female flowers.

In Situ Hybridization

A procedure utilizing nucleic corrosive hybridization with a marked test to distinguish the area of a particular mRNA in an unblemished organism.

In Vitro Fertilization

Preparation of oocytes in research facility compartments followed by counterfeit implantation of the early undeveloped organism in the mother's uterus.

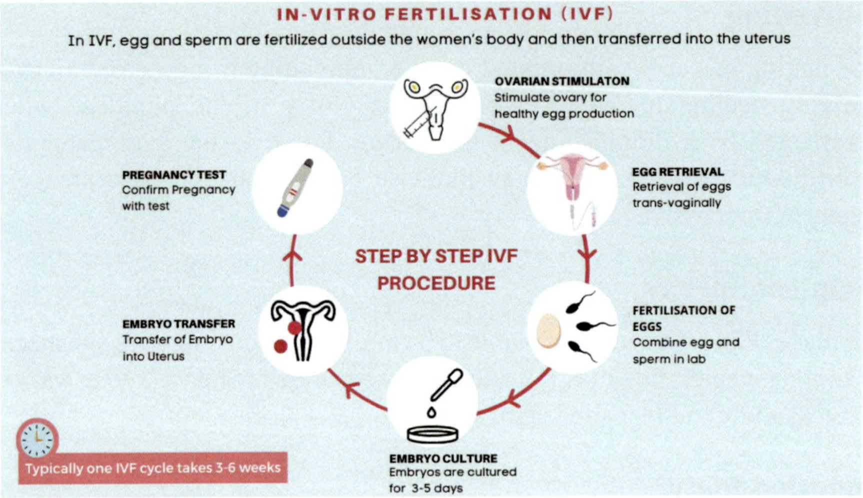

Figure 50. In Vitro Fertilization Process (IVF).

Source : Image By Wikimedia Commons

In Vitro Mutagenesis

A procedure used to mutate targeted gene in cloned DNA and then reinserting the same into the cell of an individual or organism. .

Inclusive Fitness

The ability of an individual to transfer its genes transferring to the next generation including the genes from relative family members.

Incomplete Dominance

The gene interaction wherein the heterozygotes show intermediate phenotypes.

Incomplete Flower

A flower wherein at least one of the four fundamental botanical organs (sepals, petals, stamens, or carpels) are either missing or nonfunctional.

Incomplete Metamorphosis

A kind of improvement in specific bugs, like grasshoppers, wherein the youthful (called fairies) look like grown-ups but are more modest and have different body extents. The sprite goes through a progression of sheds, each time seeming to be a grown-up until it arrives at standard size.

Indeterminate Cleavage

A kind of undeveloped improvement in deuterostomes in which every cell delivered by early cleavage divisions holds the ability to form into a total incipient organism.

Indeterminate Growth

A sort of development normal for plants, wherein the organism keeps on growing as long as it lives.

Induced Fit

It is a model where in enzyme and substrate shape their structure in such a way to bind tightly.

Inducer

A specific small molecule that binds to a bacterial repressor protein and changes the repressor's shape so that it cannot bind to an operator, thus switching an operon on.

Induction

The process in which one group of embryonic cells influences the development of another, usually by causing changes in gene expression.

Inductive Reasoning

A type of logic in which generalizations are based on a large number of specific observations.

Inflammatory Response

An innate immune defense triggered by physical injury or infection of tissue involving the release of substances that promote swelling, enhance the infiltration of white blood cells, and aid in tissue repair and destruction of invading pathogens.

Ingestion

The first stage of food processing in animals: the act of eating.

Ingroup

A species or group of species whose evolutionary relationships we seek to determine.

Inhibin

A hormone produced in the male and female gonads that functions in part by regulating the anterior pituitary by negative feedback.

Inhibitory Postsynaptic Potential (IPSP)

An electrical change (usually hyperpolarization) in the membrane of a postsynaptic neuron caused by the binding of an inhibitory neurotransmitter from a presynaptic cell to a postsynaptic receptor; makes it more difficult for a postsynaptic neuron to generate an action potential.

Innate Behavior

Animal behavior that is developmentally fixed and under strong genetic control. Innate behavior is exhibited in virtually the same form by all individuals in a population despite internal and external environmental differences during development and throughout their lifetimes.

Innate Immunity

A form of defense common to all animals that is active immediately upon exposure to pathogens and that is the same whether or not the pathogen has been encountered previously.

Inner Cell Mass

An inner cluster of cells at one end of a mammalian blastocyst that subsequently develops into the embryo proper and some of the extraembryonic membranes.

Inner Ear

One of three main regions of the vertebrate ear; includes the cochlea (which in turn contains the organ of Corti) and the semicircular canals.

Inositol Trisphosphate (Ip3)

A second messenger that functions as an intermediate between certain signaling molecules and a subsequent second messenger, Ca2+, by causing a rise in cytoplasmic Ca2+ concentration.

Inquiry

The search for information and explanation, often focusing on specific questions.

Insertion

A mutation involving the addition of one or more nucleotide pairs to a gene.

Insulin

A hormone secreted by pancreatic beta cells that lowers blood glucose levels. It promotes the uptake of glucose by most body cells and the synthesis and storage of glycogen in the liver and also stimulates protein and fat synthesis.

Integral Protein

A transmembrane protein with hydrophobic regions that extend into and often completely span the hydrophobic interior of the membrane and with hydrophilic regions in contact with the aqueous solution on one or both sides of the membrane (or lining the channel in the case of a channel protein).

Integrin

In animal cells, a transmembrane receptor protein with two subunits that interconnects the extracellular matrix and the cytoskeleton.

Integument

Layer of sporophyte tissue that contributes to the structure of an ovule of a seed plant.

Integumentary System

The outer covering of a mammal's body, including skin, hair, and nails, claws, or hooves.

Intercalated Disk

A specialized junction between cardiac muscle cells that provides direct electrical coupling between the cells.

Interferon

A protein that has antiviral or immune regulatory functions. Interferon-α and interferon-β, secreted by virus-infected cells, help nearby cells resist viral infection; interferon-γ, secreted by T cells, helps activate macrophages.

Intermediate Disturbance Hypothesis

The concept that moderate levels of disturbance can foster greater species diversity than low or high levels of disturbance.

Intermediate Filament

A component of the cytoskeleton that includes filaments intermediate in size between microtubules and microfilaments.

Internal Fertilization

The fusion of eggs and sperm within the female reproductive tract. The sperm are typically deposited in or near the tract.

Interneuron

An association neuron; a nerve cell within the central nervous system that forms synapses with sensory and/or motor neurons and integrates sensory input and motor output.

Internode

A segment of a plant stem between the points where leaves are attached.

Interphase

The period in the cell cycle when the cell is not dividing. During interphase, cellular metabolic activity is high, chromosomes and organelles are duplicated, and cell size may increase. Interphase often accounts for about 90% of the cell cycle.

Intersexual Selection

Selection whereby individuals of one sex (usually females) are choosy in selecting their mates from individuals of the other sex; also called mate choice.

Interspecific Competition

Competition for resources between individuals of two or more species when resources are in short supply.

Interspecific Interaction

A relationship between individuals of two or more species in a community.

Intertidal Zone

The shallow zone of the ocean adjacent to land and between the high- and low-tide lines.

Intracellular signaling

The communication of a molecular message from the surface of a cell to the nucleus via the participation of a series of molecules, including receptors, enzymes, proteins, and small molecules. The end result of the signaling process is the up - or down - regulation of a particular series of genes that may be involved in cell growth, division, or differentiation.

Intracytoplasmic Sperm Injection (ICSI)

The fertilization of an egg in the laboratory by the direct injection of a single sperm.

Intrasexual Selection

Selection in which there is direct competition among individuals of one sex for mates of the opposite sex.

Introduced Species

A species moved by humans, either intentionally or accidentally, from its native location to a new geographic region; also called non-native or exotic species.

Intron

A noncoding, intervening sequence within a primary transcript that is removed from the transcript during RNA processing; also refers to the region of DNA from which this sequence was transcribed.

Invasive Species

A species, often introduced by humans, that takes hold outside its native range.

Inversion

An aberration in chromosome structure resulting from reattachment of a chromosomal fragment in a reverse orientation to the chromosome from which it originated.

Invertebrate

An animal without a backbone. Invertebrates make up 95% of animal species.

Ion

An atom or group of atoms that has gained or lost one or more electrons, thus acquiring a charge.

ion channel A transmembrane protein channel that allows a specific ion to diffuse across the membrane down its concentration or electrochemical gradient.

Ion Proton II

Sequencing platform available through Gibco Life Technologies, generates 60 Gb sequence per run, with paired read lengths of 150 bp.

Ion Torrent

Sequencing platform available through Gibco Life Technologies, generates 30-120 Mb sequence per run, with paired read lengths of 400 bp.

Ionic Bond

A chemical bond resulting from the attraction between oppositely charged ions.

Ionic Compound

A compound resulting from the formation of an ionic bond; also called a salt.

Iris

The colored part of the vertebrate eye, formed by the anterior portion of the choroid.

Isomer

One of several compounds with the same molecular formula but different structures and therefore different properties. The three types of isomers are structural isomers, cis-trans isomers, and enantiomers.

Isomorphic

Referring to alternating generations in plants and certain algae in which the sporophytes and gametophytes look alike, although they differ in chromosome number.

Isoniazid

Isonicotinic acid hydrazide; an inhibitor of transamination and specifically the conversion of glycine to serine in plants. It is therefore an effective inhibitor of the photorespiratory (C2) pathway.

Isopod

A member of one of the largest groups of crustaceans, which includes terrestrial, freshwater, and marine species. Among the terrestrial isopods are the pill bugs, or wood lice.

Isoschizomers

Pairs of restriction endonucleases, obtained from different sources, that cut DNA at the same recognition site, although not necessarily at the same point in the recognition sequence.

Isotachophoresis

A specialized form of electrophoresis in which the separation is based on differences in electrophoretic mobility.

Isotonic

Descriptive of a solution of a substance in water that has the same osmotic pressure as another solution, which need not be of similar chemical composition. For instance, a glucose solution or culture medium is designed to be isotonic with a cell suspension or culture in order to prevent osmotic damage.

Isotope

Various forms of the same element having the same number of protons, but differing in the number of neutrons and hence in mass. Two types of isotopes exist: stable and unstable (radioactive). Isotopes are valuable in analytical and diagnostic work. Radioactive isotopes are detected and quantified on the basis of the ionizing radiation emitted (alpha-particles, gamma rays and beta-particles). The most useful radioactive isotopes are the low-energy beta-emitters with a reasonable half-life.

Itaconic Acid

An organic acid, produced by fermentation using certain strains of *Aspergillus niger* that are overproducers. Itaconic acid improves the properties of fibres, paints, surface coatings and carpet backings, and is used in the manufacture of adhesives.

Iteration

A series of steps in an algorithm whereby the processing of data is performed repetitively until the result exceeds a particular threshold. Iteration is often used in multiple sequence alignments whereby each set of pairwise alignments is compared with every other set, starting with the most similar pairs and progressing to the least similar, until there are no longer any sequence pairs remaining to be aligned.

Iteroparity

Reproduction in which adults produce offspring over many years; also known as repeated reproduction.

J

Java applet

A small piece of software that your browser installs automatically and that runs on your computer.

Jejunum

The region of the mammalian small intestine that lies between the ileum and the duodenum.

Jerusalem Artichoke

A plant (*Helianthus tuberosus*) related to the sunflower capable of producing high levels of carbohydrate, in the form of the fructose polymer inulin, in its underground tubers. It has been widely investigated as a potential raw material for the production of fuel alcohol (ethanol), acetone-butanol for use as a cosolvent or oxygenate in petroleum mixes, or as a source of fructose syrups.

Jet Loop Fermenter

A fermentation system in which agitation and gas absorption are achieved by recirculation of the broth with reinjection into the main fermenter vessel under pressure.

Joule (J)

A unit of energy equal to 0.28846 calories.

Juglone

A naphthoquinone derived from walnuts.

Juvenile Hormone

One of several terpenoids that inhibits metamorphosis and promotes the growth of larval structures in insects. The hormones are secreted by the corpora allata.

Juxtaglomerular Apparatus (JGA)

A specialized tissue in nephrons that releases the enzyme renin in response to a drop in blood pressure or volume.

Juxtamedullary Nephron

In mammals and birds, a nephron with a loop of Henle that extends far into the renal medulla.

K

Kanamycin

An inhibitor of protein synthesis. It acts by binding onto 70S ribosomes causing the mRNA to be misread. It is mainly used with gram-negative organisms.

Kanamycin Resistance

The ability to survive treatment with kanamycin mediated by an enzyme that modifies the antibiotic and prevents its interaction with ribosomes.

Karykinesis

The division of a cell nucleus in which the genetic complement of the daughter cells is identical to that of the parent cell.

Karyogamy

In sexual reproduction, the fusion of the nuclear material of the gametes or, especially in some fungi, the fusion of the nuclei of a heterokaryotic cell.

Karyotype

Descriptive of the chromosome complement of a cell or organism. Since each organism has a distinctive karyotype in terms of chromosome number and chromosome appearance, it represents a characteristic that can be used in the identification of a species or the parents of a hybrid, as well as in the recognition of a polyploid.

Karyotyping

The identification and classification of organisms, cells or tissues on the basis of their chromosome content.

Kefir

A fermented milk product produced in the USSR using a mixture of symbiotic yeasts, lactobacilli, leuconostocs and lactic streptococci.

Kelp

A number of different genera of large brown seaweeds, some of which are used as a source of alginic acid. Kelp has also been suggested as a source of biomass for use in anaerobic digesters for the production of biogas. Studies have been carried out growing high concentrations of kelp on floating islands.

Keratin

A structural protein with a high sulphur content that is found in hair, feathers, horn, nails, etc., as well as in the outer layer of the epidermis of vertebrates.

Kestose

A non-reducing trisaccharide formed by the condensation of sucrose and fructose with (2,6)- and (2,1)-linkages. The related trisaccharide iso-kestose is similar with (2,1):(2,1)-linkages.

Keto Acid

An acid containing a ketone group. In biological systems, organic acids with such a group in the a position are important precursors in amino acid production (e.g., a-ketoglutaric acid).

Ketogenesis

The formation of ketone bodies from acetyl-CoA produced in fatty acid oxidation.

Ketone Bodies

Compounds such as acetoacetate, hydroxybutyrate and acetone formed in the liver during fatty acid oxidation by condensation of acetyl-CoA. Ketone bodies normally pass into the blood and are transported to peripheral tissue where they are metabolized.

Ketosis

An abnormality of metabolism which results in an overproduction of ketone bodies. This may be associated with diabetes or prolonged starvation. The condition leads to acidosis and coma.

Kettle Boil

The heating of wort to inactivate enzymes, precipitate impurities and concentrate the sugar solution prior to inoculation with yeast.

Keystone Species

A species that is not necessarily abundant in a community yet exerts strong control on community structure by the nature of its ecological role or niche.

Kidney

Organ, usually paired, that performs the functions of excretion and osmoregulation in vertebrates. The individual units of the kidneys (nephrons) consist of a narrow uriniferous tubule terminating at the Bowman's capsule which encloses a knot of capillaries, the glomerulus. The capsule plus the glomerulus forms the corpuscle (malpighian body). The uriniferous tubules drain to paired ureters which open into the bladder or cloaca. The mammalian kidney is divided into an outer cortex in which the corpuscles are concentrated and an inner medulla which contains the collecting ducts. Water, salts, sugars, nitrogenous wastes and other low molecular weight compounds are filtered into the Bowman's capsule. Much of the water, together with mineral salts and glucose, is reabsorbed, and the remaining water and nitrogenous compounds are excreted as urine.

Kilobase (Kb)

A sequence of 1000 bases or base pairs in DNA, other nucleic acids or oligonucleotides. The term is used to define the size of the genome of a specific organism or of various fractions of nucleic acids following extraction or cleavage using restriction enzymes.

Kilocalorie (Kcal)

A thousand calories; the amount of heat energy required to raise the temperature of 1 kg of water by 1°C.

Kinase

An enzyme that catalyzes the phosphorylation of an organic compound; an enzyme that converts a zymogen to the active enzyme form.

Kin Selection

Natural selection that favors altruistic behavior by enhancing the reproductive success of relatives.

Kinetic Energy

The energy associated with the relative motion of objects. Moving matter can perform work by imparting motion to other matter.

Kinetic Viscosity

A description of the flow behaviour of a fluid expressed in terms of the ratio of the dynamic viscosity to the density of the material measured at the same temperature.

Kinetochore

A structure of proteins attached to the centromere that links each sister chromatid to the mitotic spindle.

Kinetoplastid

A protist, such as a trypanosome, that has a single large mitochondrion that houses an organized mass of DNA.

Kingdom

A taxonomic category, the second broadest after domain.

Kinin

(1) In animals, any of a number of polypeptides (e.g., bradykinin) that occur in blood and cause contraction of smooth muscle and dilation of blood vessels.

(2) A type of plant growth substance derived from purines that stimulates plant cell division, working synergistically with auxin. The balance between kinin and auxin also affects morphogenesis and cell expansion. Natural kinins (e.g., zeatin) are known as cytokinins or phytokinins. Synthetic analogues include kinetin (6- furfurylaminopurine) and benzyladenine (6-benzylaminopurine).

Kit

A package containing all the specialized reagents, enzymes, buffers, cloning vectors, linkers, primers, etc. required to carry out a particular diagnostic procedure or technique in molecular biology. In general, the kit provides the right proportions of the various components, packaged in a manner to make the best and most cost-effective use of expensive or difficult to obtain components. Some components may be pre-weighed, partially pre-fabricated or attached to solid supports in order to overcome the need to engage in time-consuming preparative steps.

Kjeldahl Method

An analytic technique used to determine the amount of nitrogen in a sample. It is often used for the determination of total protein in a sample on the basis of the nitrogen content. The sample is digested with concentrated sulphuric acid in the presence of potassium sulphate, copper sulphate and sodium selenate under reflux for about eight hours. The sample is then transferred quantitatively to a distillation apparatus the neck of which is immersed in a solution of boric acid. Sodium hydroxide is added to the digestion products and steam passed through the mixture so that ammonia distills over into the boric acid solution. The absorbed ammonia is then determined by titration with standard hydrochloric acid (1 ml of 0.01 M HCl corresponds to 0.14 mg of nitrogen). This method measures total nitrogen. Most proteins have a nitrogen content of about 16 percent, thus to convert the nitrogen value into a weight of protein it is multiplied by 6.25.

Klinefelter's Syndrome

A group of symptoms found in humans due to the presence of one extra X chromosome in the male karyotype (XXY).

Knockout Mice

Mice that have been engineered to lack a chosen gene. The gene is inactivated in embryonic stem cells using the technique of homologous recombination. These cells are then introduced into an early - stage embryo (blastocyst), which is then transplanted into a recipient mouse. The subsequent progeny lack the targeted gene in some cells. This technique is used to determine the function of the chosen gene.

Kojic Acid

5-Hydroxy-2-hydroxymethylgamma-pyrone; a product of glucose metabolism generated by some moulds and bacteria, including Aspergillus and Acetobacter.

Koumiss

A fermented milk product produced in the USSR. It is similar to kefir, except that it is produced from mares' milk.

K-Selection

Selection for life history traits that are sensitive to population den.

L

Lab on a Chip

A microdevice that allows rapid microanalytical analysis of DNA or protein in a single fully integrated system. Typically, these devices are miniature surfaces, made of silicon, glass, or plastic, which carry the necessary microdevices (pumps, valves, microfluidic controllers, and detectors) that allow sample separation and analysis. These devices are used in drug discovery, genetic testing, and separation science.

Label

An element, isotope or chemical compound that is attached to a biological molecule, organelle, subcellular fragment or cell in order to follow the latter's metabolism or fate. In general, labels fall into three classes: (1) isotopic labels, which may be radioactive or heavy isotopes; (2) immune labels, which may be antibodies or antigens; (3) coloured or fluorescent dyes.

Labia Majora

A pair of thick, fatty ridges that encloses and protects the rest of the vulva.

Labia Minora

A pair of slender skin folds that surrounds the openings of the vagina and urethra.

Labor

A series of strong, rhythmic contractions of the uterus that expels a baby out of the uterus and vagina during childbirth.

Laboratory Scale

Descriptive of a process carried out using only small amounts (i.e. milligrams or grams) of reagents in order to validate the nature of the products and the characteristics of the reaction.

Lac (Lactose) Operon

A transcriptional unit that codes for the three enzymes required for the metabolism of lactose by *E. coli*. In addition to the three structural genes coding for the enzymes B-galactosidase, galactoside permease and galactose acetylase, the lac operon contains a promoter site, an operator site, a site for interaction with a catabolic activator protein (CAP) and a gene coding for the synthesis of the repressor protein. These genes occur in a control region consisting of 122 nucleotides that enables the structural genes to be switched on, or induced, by a change in metabolite availability (i.e. substitution of lactose for glucose in the growth medium).

Lactation

The production of milk by the mammary glands. This is promoted by hormones produced by various glands including the pituitary, thyroid and adrenal glands, as well as by the placenta and ovaries. Secretion is stimulated by prolactin (produced by the pituitary) and adrenal corticosteroids, whereas milk ejection requires the action of oxytocin and vasopressin.

Lacteal

A tiny lymph vessel extending into the core of an intestinal villus and serving as the destination for absorbed chylomicrons.

Lactic Acid

A three-carbon organic acid produced during some fermentations and during the breakdown of glucose under anaerobic conditions.

Lactic Acid Bacteria

A trivial name applied to a diverse group of bacteria, characterized by the production of lactic acid during fermentation of carbohydrates. Genera include Lactobacillus and Streptococcus. These bacteria are of particular importance in food fermentation including dairy products (cheese, buttermilk, yoghurt and sour cream), pickles, fermented sausage and sour dough bread. Lactic acid is the main product produced by homofermentative species, whereas a mixture of lactic acid, ethanol, acetic acid, glycerol and carbon dioxide is produced by heterofermentative species.

Lactic Acid Fermentation

Glycolysis followed by the reduction of pyruvate to lactate, regenerating NAD with no release of carbon dioxide.

Lactobacillus

A genus of rod-shaped, grampositive bacteria characterized by the production of lactic acid as an end product of metabolism, through a heterofermentative or a homofermentative route. Members of this genus are widely used in the production of fermented milk products including cheese and acidophilous milk. These bacteria are also involved in beer and wine spoilage.

Laevorotatory

Descriptive of a chemical compound with optical activity. In solution, it bends polarized light to the left.

Lager

A beer produced using a bottom yeast (*Saccharomyces uvarum*). This yeast tends to flocculate and settle as the fermentation proceeds.

Lagging Strand

A discontinuously synthesized DNA strand that elongates by means of Okazaki fragments, each synthesized in a 5 3 direction away from the replication fork.

Lag Phase

The initial part of the characteristic growth curve found in most living systems. In microbial cultures, inoculation of a new medium may not result in immediate growth. Cell numbers remain stationary or increase only slowly while the cells become acclimatized to the new conditions of growth. The lag phase may be avoided if an active culture, in the exponential phase of growth, is introduced at a fairly high concentration into fresh medium of a similar composition to that used previously.

Lambda Phage

Bacteriophage; a genetically complex, but extensively studied virus of *E. coli* that has been developed as a cloning vehicle. The DNA of this phage is a linear duplex molecule of 49,000 pairs. At each end are short, single-stranded 5' -projections of 12 nucleotides, which are complementary in sequence and by which the DNA adopts a circular structure when it is injected into the host cell. Functionally related genes are grouped together. These include a central portion that is not essential for phage growth and that can be deleted and replaced without impairing the infectious growth cycle. In the lytic cycle, transcription occurs in three stages: (1) gene transcription establishes the lytic cycle; (2) gene products replicate and recombine the DNA; (3) gene products package the DNA into mature phage particles. Derivatives of the wild-type DNA have been produced that have a single target site for construction of insertional vectors, or two sites for production of replacement vectors. Hence the usual enzyme techniques of DNA cutting and rejoining can be used to insert sequences coding for other genes. The infectivity of such recombinant DNA can be increased by packaging it back into a phage particle.

Lamella

A thin, plate-like structure. In plants, lamellae are the layers of closely appressed membranes that occur in the chloroplasts. They are associated with the light-dependent biophysical and photochemical reactions. In higher plants, the lamellae generally appear in section to be stacked as grana, linked by intergranal membranes, whereas in the bundle sheath of extreme C4 plants and in most algae, such stacking is not seen. It now seems that the complex internal system of lamellae is derived from the folds and connections of a continuous sheet of membrane that divides the volume within the chloroplast envelope into two compartments, each forming a separate continuum. One compartment is the stroma proper; the other represents the intracisternal volume enclosed within the membrane. If this hypothesis is accepted, the term lamella looses much of its original meaning.

Lamina

The flattened photosynthetic blade of a leaf, or similar flattened terminal portion of the thallus of lower plants.

Laminar Flow

A description of the flow behaviour of a fluid through a pipe in which the molecules move parallel with the pipe surface. It is also known as streamline flow. In the laminar layers, transfer of momentum, heat or mass can only occur by diffusion.

Lampbrush Chromosome

A form of chromosome that occurs in the oocytes of many vertebrates. These chromosomes are larger and less condensed than the normal ones. They bear a large number of paired loops arising from the chromomere which gives it the characteristic appearance from which the name is derived.

Lancelet

Member of the clade Cephalochordata, small blade-shaped marine chordates that lack a backbone.

Landfill Gas

A mixture of methane (50-65 percent) and carbon dioxide (3{}-40 percent) together with traces of nitrogen, hydrogen and a wide range of volatile organic compounds generated by anaerobic bacteria present in landfill waste disposal sites. This gas can cause problems because of its smell, as well as being a potential hazard since it is flammable if diluted with air to a 5 - 15 percent methane content. Control measures may be needed to prevent gas migration through land drains, cracks, etc. In some cases, the gas is extracted and cleaned to provide fuel gas, which is of similar quality to natural gas. Gas production from large sites is measured in terms of thousands of cubic metres per day.

Landscape

An area containing several different ecosystems linked by exchanges of energy, materials, and organisms.

Landscape Ecology

The study of how the spatial arrangement of habitat types affects the distribution and abundance of organisms and ecosystem processes.

Large Intestine

The portion of the vertebrate alimentary canal between the small intestine and the anus; functions mainly in water absorption and the formation of feces.

Larva

A free-living, sexually immature form in some animal life cycles that may differ from the adult animal in morphology, nutrition, and habitat.

Larynx

The portion of the respiratory tract containing the vocal cords; also called the voice box.

Latent Period

The time that elapses between the reception of a stimulus and the onset of a response.

Latent Virus

A virus that remains within the host without producing any obvious effects. Activity may be induced, resulting in multiplication and the production of disease symptons, long after the initial infection.

Lateral Geniculate Nucleus

One of a pair of structures in the brain that are the destination for most of the ganglion cell axons that form the optic nerves.

Lateral Inhibition

A process that sharpens the edges and enhances the contrast of a perceived image by inhibiting receptors lateral to those that have responded to light.

Lateral Line System

A mechanoreceptor system consisting of a series of pores and receptor units along the sides of the body in fishes and aquatic amphibians; detects water movements made by the animal itself and by other moving objects.

Lateral Meristem

A meristem that thickens the roots and shoots of woody plants. The vascular cambium and cork cambium are lateral meristems.

Lateral Root

A root that arises from the pericycle of an established root.

Lateralization

Segregation of functions in the cortex of the left and right cerebral hemispheres.

Latex

A liquid found in certain plants that contains a variety of secondary products, including terpenes, terpenoids and alkaloids. High molecular weight products found in latex include rubber. Other polyisoprenes include gutta and chicle (the original chewing gum base). Latex accumulates in specific cells or vessels known as lactifers.

Lauric Acid

Dodecanoic acid; a long-chain fatty acid widely distributed in animal and vegetable fats and oils.

Law Of Conservation Of Mass

A physical law stating that matter can change form but cannot be created or destroyed. In a closed system, the mass of the system is constant.

Law Of Independent Assortment

Mendel's second law, stating that each pair of alleles segregates, or assorts, independently of each other pair during gamete formation; applies when genes for two characters are located on different pairs of homologous chromosomes or when they are far enough apart on the same chromosome to behave as though they are on different chromosomes.

Law Of Segregation

Mendel's first law, stating that the two alleles in a pair segregate (separate from each other) into different gametes during gamete formation.

Leader Peptide

A short peptide synthesized *in vitro* by translation of the leader sequence of some bacterial mRNAs. These peptides are not formed *in vivo*.

Leader Sequence

A length of mRNA, extending from the 5'-terminal to the beginning of the coding region of the first structural gene, that is not translated during protein synthesis. It may contain ribosomal binding sites or an attenuator.

Leading Strand

The new complementary DNA strand synthesized continuously along the template strand toward the replication fork in the mandatory 5 to 3 direction.

Leaf

A lateral appendage to the stem of plants. It functions in photosynthesis and transpiration. Leaves produce lateral buds in their axis. Types of true leaves include cotyledons, scale leaves, bracts and some floral parts (petals, sepals). In some plants (xerophytes), the leaves may be much reduced in size, with the leaf functions being taken over by other organs such as cladodes or phyllodes.

Leaf Primordium

A finger-like projection along the flank of a shoot apical meristem, from which a leaf arises.

Learning

The modification of behavior based on specific experiences.

Leavening Agent

A compound or system used in baked food products to produce carbon dioxide. Production of the gas causes the mix or dough to rise. Chemical leavening is carried out using baking powder (a mixture of sodium bicarbonate and acid salts). An alternative is to use a fermenting yeast culture. Baked goods prepared using fermenting yeast have a better flavour due to the reaction of fermentation products with other ingredients. However, the leavening power of yeasts is low in dough systems of high osmotic pressure. Hence breads, rolls and buns are usually leavened with yeasts, whereas cakes and biscuits are leavened using chemical agents.

Lecithin

Diacyl glyceryl phosphorylcholine or phosphatidylcholine; a type of phospholipid.

Lectins

Proteins isolated from plant tissue that are capable of binding to the antigenic glycoproteins on the surface of cells, causing the cells to agglutinate. Lectins can agglutinate red blood cells, in certain cases, with a high degree of specificity such that they may be used in typing of human blood. Lectins also preferentially agglutinate certain malignant cells. Lectins may be formed as antibodies by the plants to counteract soil bacteria or to protect against fungal attack. High concentrations of lectins are found especially in the seeds of some legumes, being localized in the cytoplasm of the cotyledons and embryonic cells. Varying levels are found in root, shoots and leaves. The lectin of soya bean has been studied in detail and the oligosaccharide moiety shown to include mannose and n-N-acetylglucosamine. Not alllectins are glycoproteins, for example, concanavalin A is a pure protein. They are also called phytoagglutinins.

Leghaemoglobin

A component of the nodules in nitrogen-fixing plants. It occurs in the bacteroid membrane, although it is synthesized on the plant genome. It is a haemoglobin-like molecule with a high affinity for oxygen and functions to protect the nitrogenase inside the bacteroid from the detrimental effects of oxygen, while providing adequate oxygen in the bacteroid membrane to allow respiration to proceed to generate the A TP required for nitrogen fixation.

Legume

(1) The trivial name for a member of the plant family Leguminosae. Members of this family are of interest since they are capable of entering a symbiotic relationship with nitrogen-fixing bacteria. (2) The fruit formed from a monocarpellary ovary in the plant family Leguminosae.

Legumin

A storage protein (along with vilivin) found in seeds of legumes. The protein has 12 components, each with an average molecular weight of 33,000. It contains more acidic and basic residues than vicilin, and three times as many tryptophan residues.

Lens

The structure in an eye that focuses light rays onto the photoreceptors.

Lenticel

A small raised area in the bark of stems and roots that enables gas exchange between living cells and the outside air. A region of loosely packed cells that occurs in the outer layer of the covering of the roots and stems (periderm) of woody plants. The cells function as a means of gaseous exchange.

Lepidoptera

Moths and butterflies; an order of insects, characterized by a covering of bright scales over the wings and bodies. They are of particular importance since the larvae (caterpillars) feed on plant leaves and stems, and are thus serious plant pests. Caterpillars are unusual in that their intestinal pH is alkaline, hence the toxic spore protein formed by the bacteria *Bacillus thuringiensis* may be used as a selective pest control agent.

Lepidosaur

Member of the reptilian group that includes lizards, snakes, and two species of New Zealand animals called tuataras.

Leptin

A hormone produced by adipose (fat) cells that acts as a satiety factor in regulating appetite.

Leptotene

A stage in the prophase of the first division of meiosis.

Lethal Gene

An allele introduced by a gene or chromosomal mutation that produces such a degree of change in an essential protein that it can no longer fulfil its proper function. If dominant, it leads to the early death of the developing embryo in all carriers. Death only occurs, however, when present in the homozygous form if it is recessive.

Leu

An abbreviation used to denote the amino acid leucine in protein sequences and elsewhere.

Leucine (Leu)

One of the twenty common amino acids found in proteins and an essential amino acid for man.

Leukocyte

A blood cell that functions in fighting infections; also called a white blood cell. A number of different types of non-pigmented, nucleate, amoeboid cells occur in the blood of animals. They are generally divided into two groups: polymorphs (basophils, neutrophils and eosinophils) characterized by a granular cytoplasm; lymphocytes and monocytes. In vertebrates, Ieucocytes protect the body against infection.

Leuconostoc

A genus of microaerophilic, spore-forming, gram-positive cocci that produce lactic acid. Leuconostoc species also produce large amounts of extracellular dextran. Species of Leuconostoc are used in the production of dextran and a number of fermented dairy products.

Leukaemia

A malignant disease, characterized by the abnormal proliferation of Ieucocytes. Some forms of leukaemia may be associated with a viral infection.

Leydig Cell

A cell that produces testosterone and other androgens and is located between the seminiferous tubules of the testes.

Lichens

Thallophytic plants formed through the symbiotic association of a fungus (mycobiont), which is either a basidiomycete or an ascomycete, and an alga (phycobiont). The morphology of the plant body is quite distinct from that of either of the constituents.

Life Cycle

The generation-to-generation sequence of stages in the reproductive history of an organism. The sum of the various stages through which an organism passes from conception to death; the description of the various stages through which an organism passes from the fertilized egg cell of one generation to the same stage in the next generation. In the simplest life cycles, the offspring is formed by simple asexual binary division. The most complex life cycles, as found in some parasitic organisms, may involve the development of a number of distinct morphological and reproductive stages. In those organisms that have sexual reproduction, an alternation of generations may be recognized, with both a haploid and a diploid genetic state; either may be dominant.

Life History

The traits that affect an organism's schedule of reproduction and survival.

Life Table

An age-specific summary of the survival pattern of a population.

Ligament

A fibrous connective tissue that joins bones together at joints. The fibrous tissue connecting bones in movable joints of vertebrates. The ligaments are made up of bundles of non-extensible collagen fibres and lie inside or outside the capsule that encloses the joint.

Ligand

A molecule that binds specifically to another molecule, usually a larger one.

Ligand-Gated Ion Channel

A transmembrane protein containing a pore that opens or closes as it changes shape in response to a signaling molecule (ligand), allowing or blocking the flow of specific ions; also called an ionotropic receptor.

Light Chain

One of the two types of polypeptide chains that make up an antibody molecule and B cell receptor; consists of a variable region, which contributes to the antigen-binding site, and a constant region.

Light Microscope (LM)

An optical instrument with lenses that refract (bend) visible light to magnify images of specimens.

Light Reactions

The first of two major stages in photosynthesis (preceding the Calvin cycle). These reactions, which occur on the thylakoid membranes of the chloroplast or on membranes of certain prokaryotes, convert solar energy to the chemical energy of ATP and NADPH, releasing oxygen in the process

Light-Harvesting Complex

A complex of proteins associated with pigment molecules (including chlorophyll a, chlorophyll b, and carotenoids) that captures light energy and transfers it to reaction-center pigments in a photosystem.

Lignin

A complex, highly ramified three dimensional polymer of phenylpropane (C0 C3) residues. In higher plants, lignin is a major constituent of the cell walls of the supporting tissues. The structural variations of the individual phenylpropane residues found in lignin are few, but the variations in the manner in which they are linked are numerous. The order of linkage is random and involves no template, hence the structure of each molecule may be unique and cannot be defined. However, it is possible to deduce a hypothetical partial structure based on the nature of the different types of building units, which on mild oxidation can release vanillin, p-hydroxybenzaldehyde and syringaldehyde. Gymnosperm lignin yields mostly vanillin and dicotyledonous lignin yields mostly vanillin and syringaldehyde, whereas monocotyledonous lignin yields all three aldehydes. The lignin molecules are probably linked covalently to polysaccharides of the cell wall matrix. The separation of lignin from other constituents (cellulose and hemicellulose) is a major problem that prevents the easy use of the very large amounts of agricultural and forestry wastes, which are generally available at low cost, as substrates for microbial growth.

Lignocellulose

The complex mixture of lignin, cellulose and hemicellulose that occurs in wood and woody parts of herbs and trees. Most intractable wastes are lignocellulose.

Lignocellulose Waste

Forestry and agricultural waste with a high lignocellulose content. These wastes represent a major source of raw material that could be exploited by application of biological techniques to produce food, animal feed, chemicals or liquid and gaseous fuels.

Limiting Factor

An environmental factor or nutritional imbalance that restricts the growth of an organism, culture or population. Although in general limiting factors may be regarded as,deficiencies, in some cases growth may be limited if the level of a given component exceeds the limits of tolerance for the organism.

Limnology

The study of freshwater ecosystems, including biological, physical and chemical aspects.

Limiting Nutrient

An element that must be added for production to increase in a particular area.

Linear Electron Flow

A route of electron flow during the light reactions of photosynthesis that involves both photosystems (I and II) and produces ATP, NADPH, and O2. The net electron flow is from H2O to NADP+.

Linkage

The occurrence of two or more genes on the same chromosome. Such linked genes do not show independent assortment, although linkage can be broken by crossing-over during meiosis. This leads to recombination (i.e. the formation of new linkage groups in the gametes and hence a new genotype in the progeny).

Linkage Analysis

The determination of the frequency of crossing-over or recombination of DNA sequences. It is used as a means of establishing the position of a gene, or sequence, in a chromosome or length of DNA.

Linkage Group

A set of gene loci that can be placed in a linear order, representing the different degrees of linkage between the loci. The term is applied to all the genes on the same chromosome, hence the number of linkage groups is equal to the haploid number of chromosomes.

Linkage Map

A genetic map based on the frequencies of recombination between markers during crossing over of homologous chromosomes.

Linked Enzyme Assay

A diagnostic procedure in which two or more enzymes are used to catalyze a series of reactions enabling identification and/or quantification of a specific substrate (or levels of a specific enzyme) and the visualization of the result. A common technique is to link one enzyme assay with a second enzyme reaction which results in either the oxidation or reduction of NAD or NADP. The rate or oxidation or reduction the coenzyme is then followed by measuring the change in absorbance at 320 nm.

Linked Genes

Genes located close enough together on a chromosome that they tend to be inherited together.

Linker

A chemical compound that is covalently bound to a length of probe DNA to enable subsequent reaction with a signal compound; a chemical compound used to join the first protected nucleotide in solid support synthesis of nucleotides to the support matrix.

Lipase

An enzyme that catalyzes the hydrolysis of fats to glycerol and fatty acids. It is more correctly described as a nonspecific acyl hydrolase that catalyzes the hydrolysis of ester bonds in monoglycerides, glycerophospholipids, diglycerides and monogalactosyl diglyceride. Commercial enzymes are produced from *Candida cylindracea*.

Lipids

One of the four major classes of compounds found in living organisms. Unlike the others (carbohydrates, proteins and nucleic acids), lipids are difficult to define in a specific way, other than by the fact that they tend to be soluble in organic solvents and insoluble in water. They include fats, waxes, phospholipids, glycolipids, steroids, terpenes and a number of different types of pigments. The major group contains those lipids whose structure is characterized by the presence of fatty acid moieties (acyl lipids). These may be divided into neutral lipids (glycerides and waxes) and polar compounds (phospholipids and glycolipids). Of the acyl lipids, glycerides accumulate as food reserves and are of commercial importance as 'fats and oils'. Phospholipids and glycolipids are structural components of membranes.

Lipolysis

The hydrolysis of lipids, and the breakdown of triglycerides in particular, to produce fatty acids which are used in oxidative catabolism. In vertebrates, hormones such as adrenaline and ACTH increase the rate of lipolysis of triglycerides in adipose tissue.

Lipopolysaccharide

A complex substance comprising a polysaccharide based on glucose, galactose, N-acetylglucosamine and other sugars linked to lipids. Lipopolysaccharides form, together with lipoproteins, the main constituents of the cell walls of gram-negative bacteria. Lipopolysaccharides are responsible for the antigenic specificity of bacteria.

Liposome

An artificial phospholipid vesicle. Macromolecules such as nucleic acids and drugs can be enclosed by liposomes.

Lipotrophin (LPH)

A product of the mammalian pituitary gland that contains peptides structurally related to the melanocyte-stimulating hormone. They stimulate lipolysis.

Liquefaction

A series of biological and/or physicochemical processes that results in the conversion of insoluble solid organic material into a liquid or soluble form. Examples of liquefaction using biocatalysts include the hydrolysis of cellulose or starch to form glucose solutions.

Liquid Chromatography

A separation technique in which a mixture of compounds in solution is passed through a column packed with a solid matrix. The sample is loaded onto the column, and the various components are eluted sequentially in the effluent from the column. Such systems may be used with packing materials for gel filtration, affinity chromatography, absorption chromatography, affinity chromatography, etc. Fully automated systems are available as high performance liquid chromatography.

Liquid Nitrogen

Nitrogen that has been liquefied using cryogenic techniques. It is used as a refrigerant to keep lyophilized stock cultures frozen at - 130°C or as a coolant for instruments such as electron microscopes.

Liver

An organ found in vertebrates that serves as the site of intermediary metabolism. It receives products of digestion absorbed from the gut via the hepatic portal vein. The liver stores carbohydrates as glycogen and also stores fats. Storage compounds are broken down and metabolized to provide energy. Other functions include the deamination of amino acids with the formation of nitrogenous excretory products, the breakdown of red blood cells, detoxification of poisonous substances and the production of bile, which is stored in the gallbladder. The liver is the site of synthesis of fibrinogen and prothrombin.

Liverwort

A small, herbaceous, nonvascular plant that is a member of the phylum Hepatophyta.

Loading

The introduction of a feedstock into a reactor or a sample for analysis into an analytical instrument, or onto a gel or similar support material in a chromatographic or electrophoretic technique.

Loam

The most fertile soil type, made up of roughly equal amounts of sand, silt, and clay.

Lobe-Fin

Member of the vertebrate clade Sarcopterygii, osteichthyans with rod-shaped muscular fins, including coelacanths, lungfishes, and tetrapods.

Local Alignment

An alignment of the most similar segments between sequences. Dissimilar sequences are not considered and are removed in the final output. BLAST does local alignments.

Local Regulator

A secreted molecule that influences cells near where it is secreted.

Locus

The region of a chromosome that is associated with the expression of a particular gene. The alleles of any one gene occupy the same loci on homologous chromosomes

Logistic Population Growth

The Population growth that levels off as population size approaches carrying capacity.

Long-Day Plant

A plant that apparently flowers in response to an extended period of light. The critical factor is in fact the period of uninterrupted darkness to which the plant is subject; a single flash of white or red light during a prolonged period of dark will stimulate flowering in long-day plants. This effect can be reversed using a flash of far-red light, indicating that a phytochrome response is involved.

Long-Term Memory

The ability to hold, associate, and recall information over one's lifetime.

Long-Term Potentiation (LTP)

An enhanced responsiveness to an action potential (nerve signal) by a receiving neuron.

Loop Fermenter

A bioreactor in which the culture medium is continuously recycled through two chambers, or through a side arm attached to a single vessel, as an aid to mixing and mass transfer.

Loop Of Henle

The hairpin turn, with a descending and ascending limb, between the proximal and distal tubules of the vertebrate kidney; functions in water and salt reabsorption.

Lophophore

In some lophotrochozoan animals, including brachiopods, a crown of ciliated tentacles that surround the mouth and function in feeding.

Lophotrochozoan

Member of a group of animal phyla identified as a clade by molecular evidence. Lophotrochozoans include organisms that have lophophores or trochophore larvae.

Low-Density Lipoprotein (LDL)

A particle in the blood made up of thousands of cholesterol molecules and other lipids bound to a protein. LDL transports cholesterol from the liver for incorporation into cell membranes.

Lowry Method

The most widely used chemical method of protein analysis. A sample containing up to 500 μg of protein is dissolved in a solution of sodium carbonate, copper sulphate, sodium hydroxide and sodium potassium tartrate, followed by addition of the phenol reagent. The dark blue colour produced, by interaction of the reagents with tyrosine in the protein, is compared colorimetrically with the colour generated using standards, usually bovine serum albumin (BSA).

Luciferin

4,5-Dihydro-2-(6-hydroxy-2- benzothiazolyl)-4-thiazolecarboxylic acid; a substrate for the light-emitting reaction responsible for the luminescent properties of many organisms. Luciferin forms a complex (luciferyl adenylate) with A TP which is then bound by the enzyme luciferase. During oxidation the activated luciferin emits light. The overall reaction is used to detect living organisms or to quantify very low levels of bacterial contamination, for instance, through the detection of the light emitted using a photomultiplier system.

Lumen

(1) The central cavity of a duct or tube. (2) The inner space of a plant cell.

Lung

An infolded respiratory surface of a terrestrial vertebrate, land snail, or spider that connects to the atmosphere by narrow tubes.

Luteal Phase

That portion of the ovarian cycle during which endocrine cells of the corpus luteum secrete female hormones.

Luteinizing Hormone (LH)

A tropic hormone that is produced and secreted by the anterior pituitary and that stimulates ovulation in females and androgen production in males; A glycoprotein secreted by the pars distalis of the pituitary, secretion being controlled by the hypothalamus. In females, the hormone promotes the maturation of the graafian follicles with the production of oestrogens and is essential for ovulation and the formation of the corpora lutea. In males, it stimulates androgen production by the interstitial cells of the testis.

Lycophyte

An informal name for a member of the phylum Lycophyta, which includes club mosses, spike mosses, and quillworts.

Lymph

The colourless fluid found in the vessels of the lymphatic system. It is derived from the interstitial fluid and contains both salts and lymphocytes, but has a lower protein content than blood plasma.

Lymph Node

A structure comprising lymphoid tissue that lies in the course of the main lymphatic vessels and acts as a filter for the removal of foreign bodies such as bacteria, thus preventing them from entering the blood stream. Lymphocytes are formed in the cortex of the lymph nodes.

Lymphatic System

A closed system of vessels through which the lymph is transported from the interstitial fluid to the blood stream. The flow of lymph in mammals is maintained by muscular activity.

Lymphocyte

A white cell arising from tissue of the lymphoid system. There are two types of lymphocytes: B cells and T cells. These cells are capable of being stimulated by an antigen to produce a specific antibody to that antigen and to proliferate to produce a population of such antibody producing cells.

Lymphoid Tissue

Tissue found in vertebrates that is responsible for the production of lymphocytes. Lymphoid tissue may be organized in the form of organs such as the lymph nodes, the thymus, spleen or tonsils, or may occur as aggregations of cells in the respiratory system and gastrointestinal tract.

Lyophilization

Freeze drying. It is used as a method of end product recovery and as a means of preserving cultures of various microorganisms. The bacteria, yeast or spores are frozen, and the water is removed by sublimation under vacuum.

Lysogenic Cycle

A type of phage replicative cycle in which the viral genome becomes incorporated into the bacterial host chromosome as a prophage, is replicated along with the chromosome, and does not kill the host.

Lysogenic Pathway

The sequence of development of a phage that results in the production of a lysogen. In this form of development, almost all of the genes carried by the phage are repressed. This requires the insertion of the phage genome into the host genome, as well as activation of repressor synthesis. The repressor represses the synthesis of the products of other genes and regulates its own synthesis, with the concentration of repressor controlling the rate of synthesis of its own mRNA.

Lysogeny

A relationship between a bacteriophage and a host bacterium in which infection does not initiate the synthesis of more phage material. Instead the phage becomes attached to a specific site on the chromosome of the bacterium, thus forming a lysogen, and is reproduced together with the chromosome being transmitted at each cell division to the daughter cells. In this prophage stage, the virus genome is repressed and the phage DNA becomes indistinguishable in behavior from a specific region of the bacterial chromosome. Spontaneous vegetative development of the phage leading to cell lysis may occur in a small proportion of lysogenic cells, releasing a number of phage particles, or this transformation can be induced by mutagens or ultraviolet radiation. Lysogenization may cause a hereditary modification of a particular bacterial property such as the ability to produce diphtheria toxins in Corynebacterium diphtheriae.

Lysosome

An organelle found in eukaryotic cells that contains hydrolytic enzymes including lipases, proteases, phosphatases and nucleases. In animals, the function of lysosomes is to digest organic matter brought into the cell by endocytosis and degenerating intracellular components. The products of digestion are released into the cytoplasm and reused. Many of the hydrolases characteristic of animal lysosomes have also been found in the vacuoles of cells from a variety of members of the plant kingdom. Hence the plant cell vacuole may be considered to function as a lysosome. Lysosomes may be derived from vesicles of the Golgi apparatus.

Lysozyme

An enzyme that catalyzes the hydrolysis of a B-(1,4)-glycosidic bond between the N-acetylmuramic acid and N-acetylglucosamine residues of the mucopeptide found in the cell walls of many bacteria. This action weakens the covalent structure of the wall and reduces its resistance to osmotic swelling, as a result of which the cells burst.

Lytic Cycle

A sequence of events that occurs when a susceptible bacterium is infected by a virulent phage. Following infection of the host by the viral chromosome, a series of enzymes are formed which, in conjunction with the pre-existing metabolic machinery of the infected cell, catalyze the formation of more phage nucleic acids and protein. In the late stage of the infection, the structural subunits are assembled within the infected cell to form new virions. At the end of this latent or vegetative phase of development, the host cell is lyzed (i.e bursts) to release the phage, which may then infect other cells and repeat the cycle. Why the lytic pathway is followed in some infected cells and the lysogenic pathway in others is not known, although the mechanism of control is known.

M

Macroevolution

Evolutionary change above the species level. Examples of macroevolutionary change include the origin of a new group of organisms through a series of speciation events and the impact of mass extinctions on the diversity of life and its subsequent recovery.

Macromolecule

A giant molecule formed by the joining of smaller molecules, usually by a dehydration reaction. Polysaccharides, proteins, and nucleic acids are macromolecules.

Macronutrient

An essential element that an organism must obtain in relatively large amounts.

Macrophage

A phagocytic cell present in many tissues that functions in innate immunity by destroying microbes and in acquired immunity as an antigen-presenting cell.

Magnoliid

Member of the angiosperm clade that is most closely related to the combined eudicot and monocot clades. Extant examples are magnolias, laurels, and black pepper plants.

Major Depressive Disorder

A mood disorder characterized by feelings of sadness, lack of self-worth, emptiness, or loss of interest in nearly all things.

Major Histocompatibility Complex (MHC) Molecule

A host protein that functions in antigen presentation. Foreign MHC molecules on transplanted tissue can trigger T cell responses that may lead to rejection of the transplant.

Malignant Tumor

A cancerous tumor containing cells that have significant genetic and cellular changes and are capable of invading and surviving in new sites. Malignant tumors can impair the functions of one or more organs.

Malpighian Tubule

A unique excretory organ of insects that empties into the digestive tract, removes nitrogenous wastes from the hemolymph, and functions in osmoregulation.

Mammal

Member of the class Mammalia, amniotes that have hair and mammary glands (glands that produce milk).

Mammary Gland

An exocrine gland that secretes milk to nourish the young. Mammary glands are characteristic of mammals.

Mantle

One of the three main parts of a mollusc; a fold of tissue that drapes over the mollusc's visceral mass and may secrete a shell. See also foot, visceral mass.

Map Unit

A unit of measurement of the distance between genes. One map unit is equivalent to a 1% recombination frequency.

Markov chain

Any multivariate probability density whose independence diagram is a chain. The variables are ordered, and each variable "depends" on its neighbours only in the sense of being conditionally independent of the others. Markov chains are an integral component of hidden Markov models.

Marsupial

A mammal, such as a koala, kangaroo, or opossum, whose young complete their embryonic development inside a maternal pouch called the marsupium.

Mast Cell

A vertebrate body cell that produces histamine and other molecules that trigger inflammation in response to infection and in allergic reactions.

Mass Extinction

The elimination of a large number of species throughout Earth, the result of global environmental changes.

Mate-Choice Copying

Behavior in which individuals in a population copy the mate choice of others, apparently as a result of social learning.

Match Score

The amount of credit given by an algorithm to an alignment for each aligned pair of identical residues.

Maternal Effect Gene

A gene that, when mutant in the mother, results in a mutant phenotype in the offspring, regardless of the offspring's genotype. Maternal effect genes, also called egg-polarity genes, were first identified in *Drosophila melanogaster*.

Matrix Genetics

A branch of bioinformatics and mathematical biology that studies the matrix forms of presentations of the genetic code.

Maximum Likelihood

As applied to molecular systematics, a principle that states that when considering multiple phylogenetic hypotheses, one should take into account the hypothesis that reflects the most likely sequence of evolutionary events, given certain rules about how DNA changes over time.

Maximum Parsimony

A principle that states that when considering multiple explanations for an observation, one should first investigate the simplest explanation that is consistent with the facts.

Mechanoreceptor

A sensory receptor that detects physical deformation in the body's environment associated with pressure, touch, stretch, motion, or sound.

Medulla Oblongata

The lowest part of the vertebrate brain, commonly called the medulla; a swelling of the hindbrain anterior to the spinal cord that controls autonomic, homeostatic functions, including breathing, heart and blood vessel activity, swallowing, digestion, and vomiting.

Medusa

The floating, flattened, mouth-down version of the cnidarian body plan. The alternate form is the polyp.

Megabase (Mb)

1,000 kilobases or 1 million bases of DNA.

Megaphyll

A leaf with a highly branched vascular system, characteristic of the vast majority of vascular plants.

Meiosis

A modified type of cell division in sexually reproducing organisms consisting of two rounds of cell division but only one round of DNA replication. It results in cells with half the number of chromosome sets as the original cell.

Meiosis I

The first division of a two-stage process of cell division in sexually reproducing organisms that results in cells with half the number of chromosome sets as the original cell.

Meiosis II

The second division of a two-stage process of cell division in sexually reproducing organisms that results in cells with half the number of chromosome sets as the original cell.

Melanocyte-Stimulating Hormone (MSH)

A hormone produced and secreted by the anterior pituitary with multiple activities, including regulating the behavior of pigment-containing cells in the skin of some vertebrates.

Melatonin

A hormone that is secreted by the pineal gland and that is involved in the regulation of biological rhythms and sleep.

Membrane Potential

The difference in electrical charge (voltage) across a cell's plasma membrane due to the differential distribution of ions. Membrane potential affects the activity of excitable cells and the transmembrane movement of all charged substances.

Memory Cell

One of a clone of long-lived lymphocytes, formed during the primary immune response, that remains in a lymphoid organ until activated by exposure to the same antigen that triggered its formation. Activated memory cells mount the secondary immune response.

Menopause

The cessation of ovulation and menstruation marking the end of a human female's reproductive years.

Menstrual Cycle

In humans and certain other primates, a type of reproductive cycle in which the nonpregnant endometrium is shed through the cervix into the vagina; also called the uterine cycle.

Meristem

Plant tissue that remains embryonic as long as the plant lives, allowing for indeterminate growth.

Meroblastic

Referring to a type of cleavage in which there is incomplete division of a yolk-rich egg, characteristic of avian development.

Mesoderm

The middle primary germ layer in a triploblastic animal embryo; develops into the notochord, the lining of the coelom, muscles, skeleton, gonads, kidneys, and most of the circulatory system in species that have these structures.

Mesophyll

Leaf cells specialized for photosynthesis. In C3 and CAM plants, mesophyll cells are located between the upper and lower epidermis; in C4 plants, they are located between the bundle-sheath cells and the epidermis.

Messenger RNA (mRNA)

A type of RNA, synthesized using a DNA template, that attaches to ribosomes in the cytoplasm and specifies the primary structure of a protein. (In eukaryotes, the primary RNA transcript must undergo RNA processing to become mRNA.)

Metabolic Pathway

A series of chemical reactions that either builds a complex molecule (anabolic pathway) or breaks down a complex molecule to simpler molecules (catabolic pathway).

Metabolism

The totality of an organism's chemical reactions, consisting of catabolic and anabolic pathways, which manage the material and energy resources of the organism.

Metagenomics

The collection and sequencing of DNA from a group of species, usually an environmental sample of microorganisms. Computer software sorts partial sequences and assembles them into genome sequences of individual species making up the sample.

Metamorphosis

A developmental transformation that turns an animal larva into either an adult or an adult-like stage that is not yet sexually mature.

Metanephridium

An excretory organ found in many invertebrates that typically consists of tubules connecting ciliated internal openings to external openings.

Metaphase

The third stage of mitosis, in which the spindle is complete and the chromosomes, attached to microtubules at their kinetochores, are all aligned at the metaphase plate.

Metaphase Plate

An imaginary structure located at a plane midway between the two poles of a cell in metaphase on which the centromeres of all the duplicated chromosomes are located.

Metapopulation

A group of spatially separated populations of one species that interact through immigration and emigration.

Methanogen

An organism that produces methane as a waste product of the way it obtains energy. All known methanogens are in domain Archaea.

Methyl Group

A chemical group consisting of a carbon bonded to three hydrogen atoms. The methyl group may be attached to a carbon or to a different atom.

Microarray

Technology used for parallel gene expression and DNA homology analysis using an orderly arrangement of thousands of identified sequenced genes printed on a solid chip

Microevolution

Evolutionary change below the species level; change in the allele frequencies in a population over generations.

Microfilament

A cable composed of actin proteins in the cytoplasm of almost every eukaryotic cell, making up part of the cytoskeleton and acting alone or with myosin to cause cell contraction; also known as an actin filament.

Microrna (Mirna)

A small, single-stranded RNA molecule, generated from a hairpin structure on a precursor RNA transcribed from a particular gene. The miRNA associates with one or more proteins in a complex that can degrade or prevent translation of an mRNA with a complementary sequence.

Microtubule

A hollow rod composed of tubulin proteins that makes up part of the cytoskeleton in all eukaryotic cells and is found in cilia and flagella.

Microvillus

One of many fine, finger-like projections of the epithelial cells in the lumen of the small intestine that increase its surface area.

Midbrain

One of three genealogical and early stage districts of the vertebrate mind; forms into tangible incorporating and transfer focuses that send tactile data to the frontal cortex.

Middle Ear

One of three fundamental districts of the vertebrate ear; in well-evolved creatures, is a chamber containing three little bones (the malleus, incus, and stapes) that pass vibrations from the eardrum onto the oval window.

Middle Lamella

In plants, a slender layer of glue extracellular material, fundamentally gelatins, is found between the essential dividers of neighbouring youthful cells.

MIM Number

Also known as MIM#, OMIM number, or McKusick code. A unique six - digit number assigned to each entry listed in the catalog of human genes and genetic disorders, “Online Mendelian Inheritance in Man ” (OMIM).

MiSeq

Illumina sequencing platform that generates 15 Gb sequence per run with paired read lengths of 300 bp.

Mismatch Repair

The cell interaction utilises explicit compounds to eliminate and supplant inaccurately matched nucleotides.

Mismatch Score

The penalty assigned by an algorithm when nonidentical restudies are aligned in an alignment.

Missense Mutation

A nucleotide-pair replacement that outcomes in a codon that codes for an alternate amino corrosive.

Mitochondrial Matrix

The compartment of the mitochondrion is encased by the internal layer and contains chemicals and substrates for the citrus extract cycle, as well as ribosomes and DNA.

Mitochondrial Signal Sequence

A string of amino acids that causes a eukaryotic protein to be delivered to a cell ' s mitochondria.

Mitochondrion

An organelle in eukaryotic cells that fills in as the site of cell breath; utilises oxygen to separate natural particles and orchestrate ATP.

Mitosis

A course of atomic division in eukaryotic cells is traditionally partitioned into five phases: prophase, prometaphase, metaphase, anaphase, and telophase. Mitosis moderate's chromosome number by distributing reproduced chromosomes similarly to every one of the little girl cores.

Mitotic Spindle

A collection of microtubules and related proteins that is associated with the development of chromosomes during mitosis.

Modeling

Refers to molecular modeling, a process whereby the three - dimensional architecture of biological molecules is interpreted (or predicted), visually represented, and manipulated in order to determine their molecular properties. (general) a series of mathematical equations or procedures that simulate a real - life process given a set of assumptions, boundary parameters, and initial conditions.

Model Organism

Specific animal categories were picked for an investigation into expansive natural standards since it is illustrative of a bigger gathering and generally simple to fill in a lab.

Molarity

A typical proportion of solute fixation, alluding to the number of moles of solute per litre of arrangement.

Mould

Casual expression for a parasite that develops as a filamentous organism, delivering haploid spores by mitosis and shaping a noticeable mycelium.

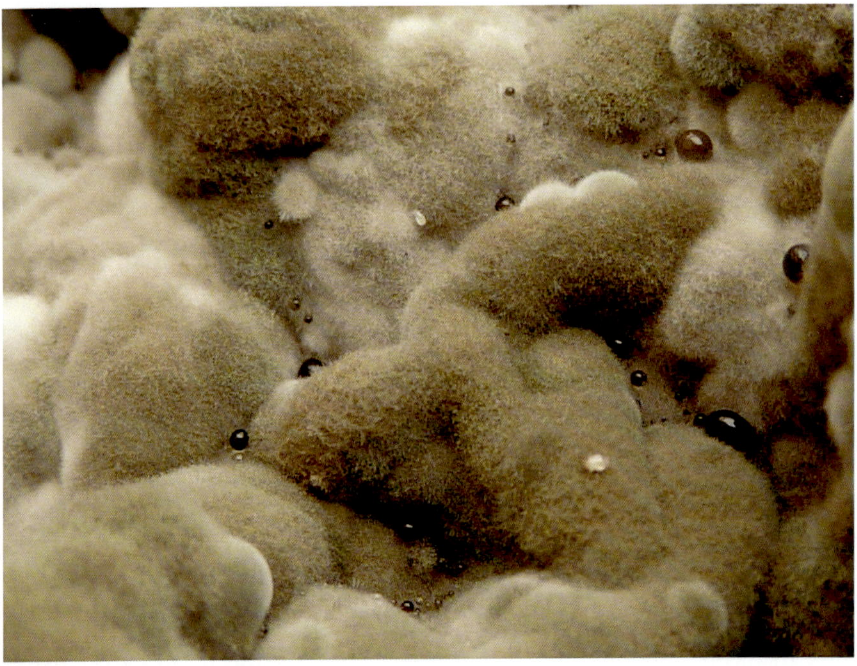

Figure 51. Mold Landscape.

Source : Image By Flickr

Mole (Mol)

The number of grams of a substance equivalents its sub-atomic load in daltons and contains Avogadro's number of particles.

Molecular Clock

A technique for assessing the time expected for a given measure of transformative change, in light of the perception that a few districts of genomes develop at consistent rates.

Molecular function

Coined by Gene Ontology, describes tasks performed by individual gene products such as transcription factor or calcium transportation.

Molecular Systematics

A logical discipline that utilizes nucleic acids or different particles to construe transformative connections between various species.

Monoclonal Antibody

Any arrangement of antibodies that have been delivered by a solitary clone of refined cells and along these lines are explicit for a similar epitope.

Monohybrid

An organism that is heterozygous concerning a single gene of interest. All the progenies from a cross between homozygous parent for various alleles is monohybrids. For instance, Parent of genotypes AA and aa produce a monohybrid of genotype Aa.

Monohybrid Cross

A cross between two individual's homozygous for single gene.

Monomer

A single unit of any biological molecule or macromolecule, such as an amino acid, nucleic acid, polypeptide domain, or protein.

Monophyletic

Relating to a gathering of taxa that comprises a typical predecessor and the entirety of its relatives. A monophyletic taxon is comparable to a clade.

Monosaccharide

The most straightforward starch, dynamic alone or filling in as a monomer for disaccharides and polysaccharides. Otherwise called straightforward sugars, monosaccharides have sub-atomic equations that are for the most part different from CH_2O.

Monosomic

Alluding to a diploid cell that has just a single complement of a specific chromosome rather than the ordinary two.

Monotreme

An egg-laying warm-blooded animal, like a platypus or echidna. Like all well-evolved animals, monotremes have hair and produce milk, however, they need areolas.

Morphogen

A substance, for example, Bicoid protein in Drosophila, that gives positional data as a fixation inclination along with an early stage pivot.

Motif

Short recurring patterns in DNA that are thought to have a biological function.

Motor Protein

A protein that communicates with cytoskeletal components and other cell parts. Motor proteins convert chemical energy into mechanical energy through hydrolysis of ATP.

Movement System

An efferent part of the vertebrate fringe sensory system is made out of engine neurons that convey signs to skeletal muscles in light of outside improvements.

MPF

Development advancing component (or M-stage advancing variable); a protein complex expected for a cell to advance from late interphase to mitosis. The dynamic structure comprises cyclin and a protein kinase.

Mucus

A thick and dangerous combination of glycoproteins, cells, salts, and water that saturates and safeguards the layers lining body depressions that open to the outside.

Multiplication Rule

A standard of likelihood expresses that the likelihood of at least two autonomous occasions happening together is not entirely set in stone by duplicating their singular probabilities.

Muscle Tissue

Tissue comprising of long muscle cells that can contract, either all alone or when invigorated by nerve driving forces.

Myelin Sheath

Folded over the axon of a neuron, a protecting layer of cell films from Schwann cells or oligodendrocytes. It is interfered with by hubs of Ranvier, where activity possibilities are produced.

Myofibril

A longitudinal group in a muscle cell (fibre) that contains slender fibres of actin and administrative proteins and thick fibres of myosin.

Myosin

A kind of engine protein that partners with fibres that communicate with actin fibres to cause cell constriction.

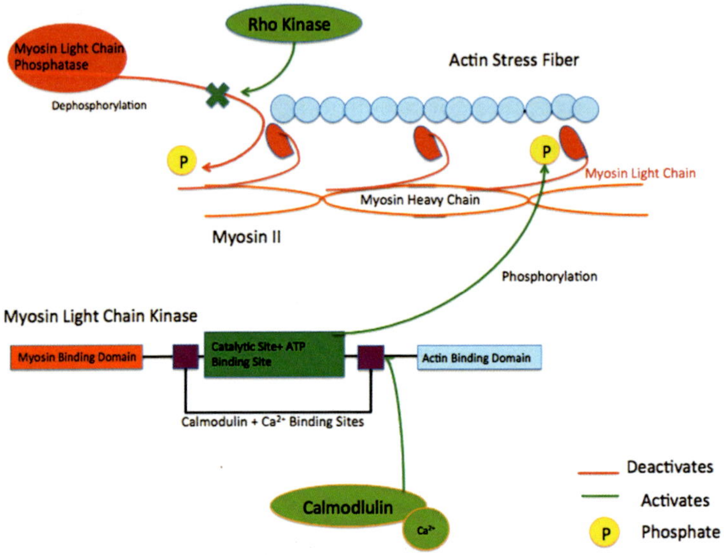

Figure 52. Image Shows Myosin Light Chain Kinase Protein Allosterically Activated By Calmodulin.

Source : Image By Wikimedia Commons

Myriapod

An earthbound arthropod with many body fragments and a couple of sets of legs per section. Millipedes and centipedes are the two significant gatherings of living myriapods.

N

NAD+

Nicotinamide adenine dinucleotide, a coenzyme that cycles effectively between oxidised (NAD) and diminished (NADH) states, along these lines going about as an electron transporter.

NADP+

Nicotinamide adenine dinucleotide phosphate is an electron acceptor that, as NADPH, briefly stores invigorated electrons delivered during the light responses.

Natural Family Planning

A type of contraception that depends on shunning sex when origination is probably going to happen; is additionally called the musicality technique.

Natural Killer Cell

A sort of white platelet that can kill cancer cells and infection contaminated cells as a component of intrinsic insusceptibility.

Natural Selection

An interaction wherein people that have specific acquired characteristics will quite often make due and duplicate at higher rates than others as a result of those attributes.

NCBI (National Centre for Biotechnology Information)

A federally funded part of the U.S. National Library of Medicine. NCBI is the home of GenBank, BLAST, COG, and many other genomic databases and computational tools.

Negative Feedback

A type of guideline wherein aggregation of a finished result of an interaction eases back the cycle; in physiology, an essential instrument of homeostasis, by which an adjustment of a variable triggers a reaction that neutralises the underlying change.

Nematocyst

In a cnidocyte of a cnidarian, a container like an organelle containing a snaked string that when released can infiltrate the body mass of the prey.

Nervous System

The effective interior arrangement of correspondence includes tangible receptors, organisations of nerve cells, and associations with muscles and organs that answer nerve signals; capacities working together with the endocrine framework to impact inside guidelines and keep up with homeostasis.

Net Ecosystem Production (NEP)

The gross essential creation of a biological system shortens the energy utilised by all autotrophs and heterotrophs for breath.

Net Primary Production (NPP)

The gross essential creation of a biological system shortens the energy involved by the makers for breath.

Neural Crest

In vertebrates, a district is situated at the edges of the drain tube where it squeezes off from the ectoderm. Brain peak cells relocate to different pieces of the undeveloped organism and structure shade cells in the skin and portions of the skull, teeth, adrenal organs, and fringe sensory system.

Neural Tube

A container of infolded ectodermal cells that runs along with the foremost back pivot of a vertebrate, only dorsal to the notochord. It will bring about the focal sensory system.

Neurohormone

A particle that is discharged by a neuron, goes into body liquids, and follows up on unambiguous objective cells, changing their working.

Neuron

A nerve cell; is the principal unit of the sensory system, and has construction and properties that permit it to direct announces exploiting the electrical charge across its plasma layer.

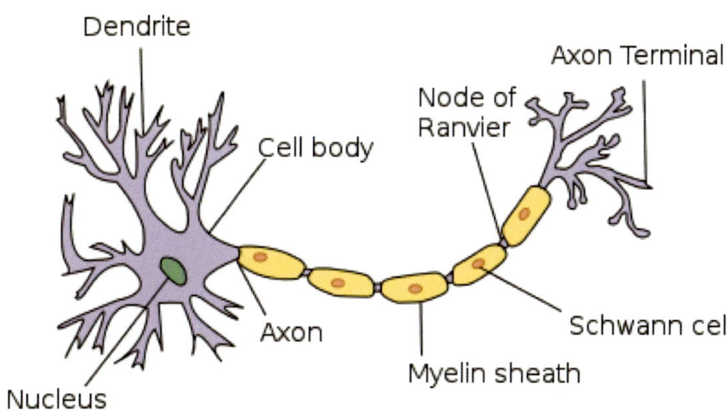

Figure 53. Diagram Of A Neuron.

Source : Image By Wikimedia Commons

Neurotransmitter

A particle that is let out of the synaptic terminal of a neuron at a synthetic neurotransmitter, diffuses across the synaptic separate, and ties to the postsynaptic cell, setting off a reaction.

Neutral Theory

The speculation that much transformative change in qualities and proteins significantly affects the wellness and along these lines isn't impacted by regular determination.

Neutron

A subatomic molecule having no electrical charge (electrically nonpartisan), with a mass of around 1.7 * 10-24 g, is found in the core of a particle.

Neutrophil

The most bountiful sort of white platelet. Neutrophils are phagocytic and keep an eye on fall to pieces as they obliterate unfamiliar trespassers, restricting their life expectancy to a couple of days.

NextSeq

Illumina sequencing platform that generates 120 Gb sequence per run with paired read lengths of 150 bp.

Nitric Oxide (NO)

A gas delivered by many sorts of cells that capacities as a neighbourhood controller and as a synapse.

Nitrogen Cycle

The regular interaction by which nitrogen, either from the environment or from a disintegrated natural material, is changed over by soil microscopic organisms to compounds acclimatised by plants. This consolidated nitrogen is then taken in by different organic entities and along these lines delivered, followed up on by microorganisms, and made accessible again to the nonliving climate.

Nitrogen Fixation

The transformation of environmental nitrogen (N2) to smelling salts (NH3). Natural nitrogen obsession is completed by specific prokaryotes, some of which have mutualistic associations with plants.

Nociceptor

A tangible receptor that answers poisonous or excruciating improvements; additionally, called a torment receptor.

Node Of Ranvier

Hole in the myelin sheath of specific axons where an activity potential might be produced. In saltatory conduction, an activity potential is recovered at every hub, seeming to "bounce" along the axon from one hub to another.

Nodule

An enlarging on the foundation of a vegetable. Knobs are made out of plant cells that contain nitrogen-fixing microscopic organisms of the variety Rhizobium.

Non-Competitive Inhibitor

A substance that diminishes the movement of a catalyst by restricting to an area remote from the dynamic site, changing the compound's shape so the dynamic site never again successfully catalyses the transformation of the substrate to item.

Nondisjunction

A mistake in meiosis or mitosis wherein individuals from a couple of homologous chromosomes or a couple of sister chromatids neglects to isolate appropriately from one another.

Nonequilibrium Model

A model that keeps up with that network changes continually in the wake of being slammed by unsettling influences.

Nonpolar Covalent Bond

A sort of covalent bond wherein electrons is shared similarly between two particles of comparative electronegativity.

Nonsense Mutation

A transformation that changes an amino corrosive codon to one of the three stop codons, bringing about a more limited and typically nonfunctional protein.

Norepinephrine

A catecholamine that is synthetically and practically like epinephrine and goes about as a chemical or synapse; otherwise called noradrenaline.

Normal Distribution

Data have an average value around which all the individuals are clustered. A graph of normal distribution results in the classic 'bell curve'.

Normalized Data

Data that has been corrected or standardized, for example, by subtracting the sample mean from each observation, and dividing the result by the sample standard deviation.

Northern Blotting

A procedure that empowers explicit nucleotide arrangements to be distinguished in examples of mRNA. It includes gel electrophoresis of RNA particles and their exchange to a layer (blotching), trailed by nucleic corrosive hybridisation with a marked test.

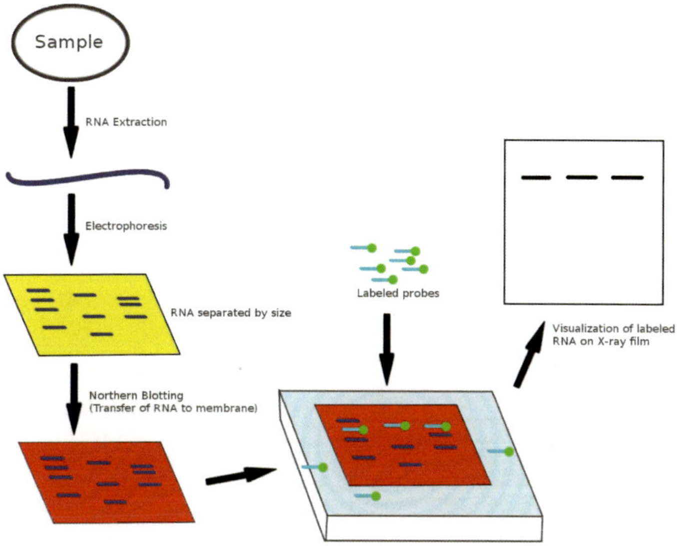

Figure 54. Northern Blot Diagram.

Source : Image By Wikimedia Commons

Notochord

A longitudinal, adaptable pole made of firmly pressed mesodermal cells that runs along with the foremost back pivot of a chordate in the dorsal piece of the body.

Nuclear Envelope

In a eukaryotic cell, the twofold layer that encompasses the core is punctured with pores that control traffic with the cytoplasm. The external layer is persistent with the endoplasmic reticulum.

Nuclear Lamina

A netlike cluster of protein fibres that lines the internal surface of the atomic envelope and keeps up with the state of the core.

Nucleariid

Individuals from a gathering of unicellular, amoeboid protists that are more firmly connected with organisms than they are to different protists.

Nuclease

A chemical that cuts DNA or RNA, either eliminating one or a couple of bases or hydrolysing the DNA or RNA totally into its part nucleotides.

Nucleic Acid

A polymer (polynucleotide) comprising numerous nucleotide monomers; fills in as a diagram for proteins and, through the activities of proteins, for every single cell movement. The two sorts are DNA and RNA.

Nucleic Acid Hybridization

The course of base matching between a quality and a corresponding arrangement on another nucleic corrosive particle.

Nucleic Acid Probe

In DNA innovation, a marked single-abandoned nucleic corrosive particle is used to find a particular nucleotide arrangement in a nucleic corrosive example. Particles of the test hydrogen attach to the corresponding arrangement any place it happens; radioactive, fluorescent, or other markings of the test permit its area to be distinguished.

Nucleolus

The fundamental, beadlike unit of DNA pressing in eukaryotes, comprising of a fragment of DNA twisted around a protein center made out of two duplicates of every one of four kinds of histone.

Nucleotide

The structured square of a nucleic corrosive, comprising of a five-carbon sugar covalently attached to a nitrogenous base and at least one phosphate gathering.

Nucleotide Excision Repair

A maintenance framework that eliminates and afterwards accurately replaces a harmed fragment of DNA involving the unharmed strand as an aide.

Nucleotide-Pair Substitution

A kind of point transformation wherein one nucleotide in a DNA strand and its accomplice in the corresponding strand are supplanted by one more set of nucleotides.

Nucleus

(1) profoundly, containing protons and neutrons.

(2) The organelle of a eukaryotic cell that contains the hereditary material as chromosomes, comprised of chromatin.

(3) A bunch of neurons.

O

Obligate Anaerobe

An organism that primarily completes its lifecycle in anaerobic environment. Such animals can't utilise oxygen and as a matter of fact, might be harmed by it.

Object - Relational Database

Databases that combine the elements of object orientation and object - oriented programming languages with database capabilities. They provide more than persistent storage of programming language objects. Object databases extend the functionality of object programming languages (e.g., C ++, Smalltalk, Java) to provide full - featured database programming capability. The result is a high level of congruence between the data model for the application and the data model of the database.

Ocean Acidification

Diminishing pH of sea waters because of assimilation of an overabundance of environmental CO2 from the consumption of non-renewable energy sources.

Okazaki Fragment

A short fragment of DNA is orchestrated away from the replication fork on a layout strand during DNA replication. Many such fragments are consolidated to make up the slacking strand of recently orchestrated DNA.

Oligodendrocyte

A kind of glial cell that structures protecting myelin sheaths around the axons of neurons in the focal sensory system.

Oncogene

A quality found in viral or cell genomes that are associated with setting off sub-atomic occasions that can prompt malignant growth.

Oogenesis

The interaction in the ovary outcomes in the development of female gametes.

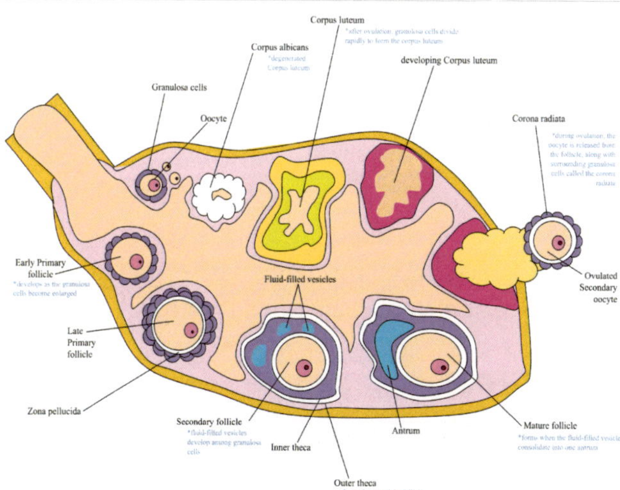

Figure 55. Image Showing Oogenesis Process.

Source : Image By Wikimedia Commons

OOMYCETE

A protist with flogged cells, for example, a water shape, white rust, or fleece buildup, that procures nourishment for the most part as a decomposer or plant parasite.

Open Circulatory System

A circulatory framework wherein liquid called hemolymph washes the tissues and organs straightforwardly and there is no differentiation between the circling liquid and the interstitial liquid.

Operator

In bacterial and phage DNA, an arrangement of nucleotides close to the beginning of an operon to which a functioning repressor can connect. The limiting of the repressor keeps RNA polymerase from connecting to the promoter and interpreting the function of the operon.

Operon

A unit of hereditary capacity found in microscopic organisms and phages, comprising of an promotor, an operator, and a coordinately controlled group of genes whose items work in a typical pathway.

Opisthokont

An individual from the different clade Opisthokonta, organisms that plummeted from a predecessor with a back flagellum, including microbes, animals, and certain protists.

Optic Chiasm

Where the two optic nerves meet and axons addressing unmistakable sides of the visual field are isolated from each other prior to arriving at the cerebrum.

Organ Identity Gene

A plant homeotic quality that utilises positional data to specify four floral organs; sepals, petals, stamen and carpel.

Organ Of Corti

The genuine hearing organ of the vertebrate ear, situated on the floor of the cochlear conduit in the internal ear; contains the receptor cells (hair cells) of the ear.

Organelle

Any of a few layer encased structures with particular role, present in the cytosol of eukaryotic cells.

Organismal Ecology

The part of biology worried about the morphological, physiological, and conduct manners by which individual organic entities address the difficulties presented by their biotic and abiotic conditions.

Organogenesis

The interaction wherein organ fundamentals create from the three microorganism layers after gastrulation.

Orgasm

Cadenced, compulsory constrictions of specific regenerative designs in the two genders during the human sexual reaction cycle.

Origin Of Replication

The site where the replication of a DNA particle starts, comprising of a specific sequence of nucleotides.

Outer Ear

One of three fundamental area of the ear in reptiles (counting birds) and warm-blooded animals; comprised of the hear-able waterway and, in many birds and vertebrates, the pinna.

Outgroup

Group of distantly related organisms that serve as reference to study the evolutionary relationship of ingroup organisms.

Oval Window

In the vertebrate ear, a layer canvassed hole in the skull bone, through which sound waves pass from the centre ear to the inner ear. Also known as fenestra ovalis.

Ovarian Cycle

The cyclic repeat of the follicular stage, ovulation, and the luteal stage in the mammalian ovary is controlled by hormones.

Ovary

(1) In flowers, the piece of a carpel in which the egg-containing ovules create.

(2) In animals, the organ produces female gametes and regenerative hormones.

Oviduct

A cylinder passes from the ovary to the vagina in spineless animals or to the uterus in vertebrates, where it is otherwise called a fallopian tube.

Oviparous

Organism's that produces young from egg hatched by mother.

Ovoviviparous

Alluding to a producing young animals directly from eggs that are hatched in the body of mother.

Ovulation

The release of an egg from an ovary. In females, an ovarian follicle release an egg during each uterine/feminine cycle.

Ovule

A structure that creates inside the ovary of a seed plant and contains the female gametophyte.

Oxidation

The total or incomplete loss of electrons from a substance is associated with a redox response.

Oxidative Phosphorylation

The development of ATP utilising energy got from the redox responses of an electron transport chain; the third significant phase of cell respiration process.

Oxytocin

A hormone that is produced by the hypothalamous and released from the posterior pituitary. It instigates constrictions of the uterine muscles during work and makes the mammary organs discharge milk during nursing.

P

P Site

One of a ribosome's three restricting destinations for tRNA during translation. The P site holds the tRNA carrying the developing polypeptide chain. (P represents peptidyl tRNA.)

P53 Gene

A tumour suppressor gene that regualtes the cell cycle and culminate the tumour formation process. The gene is reported to be located on Chromosome 17.

Paedomorphosis

The retention of larval or juvenile character in the adult stage of the organism.

Pancreas

An organ with exocrine and endocrine tissues. The exocrine piece capacities in assimilation, discharging chemicals and an antacid arrangement into the small digestive system by means of a channel; the ductless endocrine part works in homeostasis, emitting the chemicals insulin and glucagon into the blood.

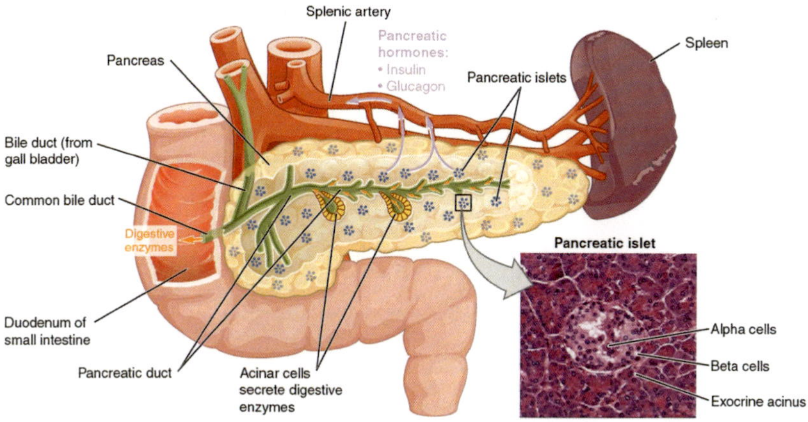

Figure 56. The Pancreas.

Source : Image By Wikimedia Commons

Pangaea

The supercontinent that created by the end of Paleozoic age or beginning of Mesozoic age when plate developments brought every one of the land massess of Earth together.

Parameters

Selectable values, typically determined experimentally, that govern the boundaries of an algorithm or program. For example, selection of the appropriate input parameters governs the success of a search algorithm. Some of the most common search parameters in bioinformatics tools include the stringency of an alignment search tool and the weights (penalties) provided for mismatches and gaps.

Paraphyletic

Relating to a grouping of taxa that comprises a typical predecessor and some, however not all, of its relatives.

Parareptile

A clade of basal sauropsids, comprising for the most part of enormous, stocky quadrupedal herbivores. Parareptiles ceased to exist in the late Triassic time frame.

Parasite

An organism that feeds on the cell items, tissues, or body liquids of another species (the host) while in or on the host organism. Parasites damages host however for the most part don't kill their host.

Parasitism

An advantageous relationship wherein one organism, the parasite, benefits to the detriment of another, the host, by residing either inside or on the host.

Parasympathetic Division

One of three divisions of the autonomic sensory system; for the most part improves body exercises that increase and preserve energy, for example, assimilation and diminished pulse.

Parathyroid Gland

One of four little endocrine organs, implanted in the outer layer of the thyroid organ, that discharges parathyroid chemicals.

Parathyroid Hormone (PTH)

A chemical discharged by the parathyroid organs raises blood calcium levels by advancing calcium discharge from bone and calcium maintenance by the kidneys.

Parenchyma Cell

A moderately unspecialised plant cell type that completes the vast majority of the digestion, orchestrates and stores natural items, and forms into a more separated cell type.

Parkinson's Disease

An ever-evolving cerebrum infection is described by trouble in starting developments, gradualness of development, and unbending nature.

Parthenogenesis

A type of asexual reproduction wherein females produces progeny from unfertilised eggs.

Partial Pressure

The tension is applied by a specific gas in a combination of gases (for example, the strain applied by oxygen in the air).

Passive Immunity

The transient insusceptibility is presented by the exchange of antibodies, as happens in the exchange of maternal antibodies to a hatchling or nursing newborn child.

Pathways

Bioinformatics strives to define representations of key biological data types, algorithms, and inference procedures, including sequences, structures, biological pathways, and reactions. Representing and computing with biological pathways requires ontologies for representing pathway knowledge, user interfaces to these databases, physicochemical properties of enzymes and their substrates in pathways, and pathway analysis of whole genomes, including identifying common patterns across species and species differences.

Pattern Formation

The improvement of a multicellular organism's spatial association, the course of action of organs and tissues in their trademark places in a three-layered space.

Peat

Partially decomposed material primarily found in wetland area consisting plant residual organic matter.

Pedigree

An outline of a genealogical record with traditional images, showing the transmission of heritable characters in parents and ancestors over different ages.

Pep Carboxylase

Phosphoenolpyruvate carboxylase ia a enzyme that adds CO2 to phosphoenolpyruvate (PEP) to form oxaloacetate in mesophyll cells of C4 plants in Hatch Slack pathway. This reaction is first step in photosynthesis in C4 plant.

Peptide Bond

The covalent bond between the carboxyl group on one amino acid and the amino group on another is framed by a lack of hydration response.

Peptidoglycan

A type of polymer in bacterial cell membranes comprising of sugars interconnected by short polypeptides.

Periderm

The defensive coat that replaces the epidermis in woody plants during auxiliary development, is formed by the endless cork cambium.

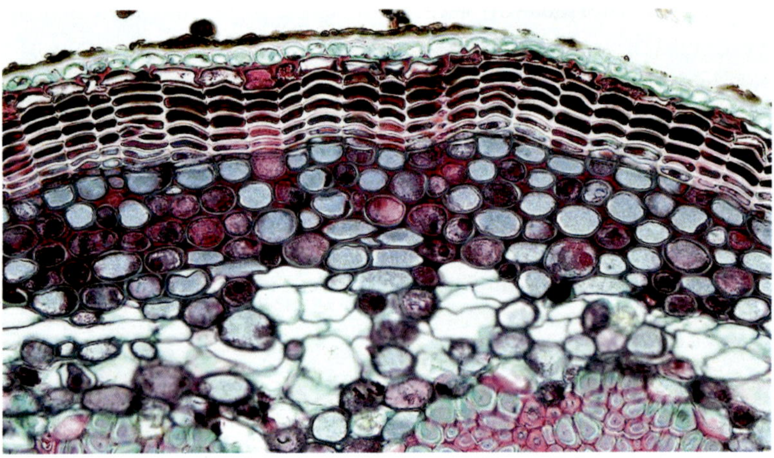

Figure 57. Image Showing Periderm.

Source : Image By Wikimedia Commons

Peripheral Protein

A protein is firmly bound to the integral membrane protein and attach to the the lipid bilayer.

Peristalsis

(1) Alternating influxes of constriction and unwinding in the smooth muscles covering the nutritious waterway that move food along the channel.

(2) A kind of development ashore delivered by cadenced influxes of muscle constrictions passing from front to back, as in numerous annelids.

Peristome

In certain Mosses, it is a ring of interlocking, tooth-like designs on the upper side (sporangium), generally involved in spore release.

Peritubular Capillary

One of the minuscule vessels that encompassing the cortical parts of proximal and distal tubules in the kidney.

Peroxisome

A membrane bound organelle containing compounds that move hydrogen particles from different substrates to oxygen (O2), delivering and afterwards altering the hydrogen peroxide (H2O2).

Petal

A modified leaf of a flowering plant. Petals are the segment of corrola that attracts it to bugs and different pollinators.

Phagocytosis

A type of endocytosis in which enormous particulate substances or little organic substances are taken up by a cell. It is cfound in certain protists and by specific insusceptible cells of animals (in warm-blooded animals, for the most part, macrophages, neutrophils, and dendritic cells).

Pharyngeal Cleft

In chordate undeveloped organisms, one of the depressions that different a progression of pockets at the edges of the pharynx and may form into a pharyngeal cut.

Pharyngeal Slit

In chordate undeveloped organisms, one of the cuts that structure from the pharyngeal clefts and impart to the outside, later forming into gill cuts in many vertebrates.

Pharynx

1) A region in the vertebrate throat where air and food sections cross.

(2) In flatworms, the strong cylinder projects from the ventral side of the worm and ends in the mouth.

Phenotype

The recognisable physical and physiological characteristics of an organism are not entirely due its hereditary design.

Pheromone

In animals and parasites, a little particle is delivered into the climate that helps in correspondence between individuals from similar species. In animals, it acts similar to a chemical in impacting physiology and behaviour.

Phloem

Vascular plant tissue comprising of living cells organised into lengthened tubes that transport sugar and other natural supplements all through the plant.

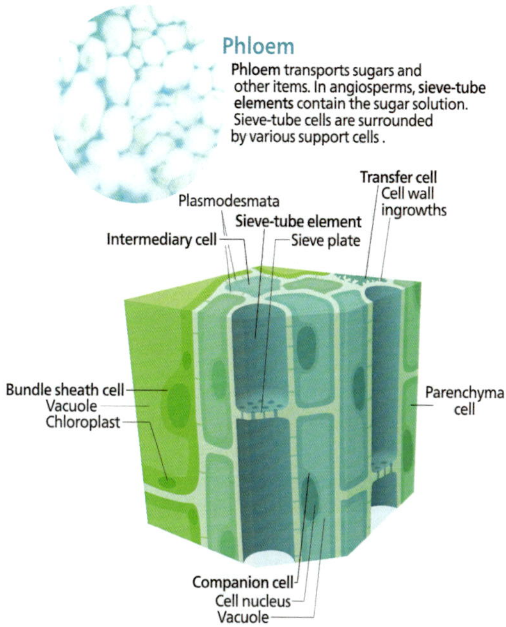

Figure 58. Phloem Tissue In Plants.

Source : Image By Wikimedia Commons

Phosphate Group

A synthetic group comprising of a phosphorus particle attached to four oxygen molecules; significant in energy transport.

Phospholipid

A lipid comprised of glycerol connected to two unsaturated fats and a phosphate group. The hydrocarbon chains of the unsaturated fats go about as nonpolar, hydrophobic tails, while the remainder of the particle goes about as a polar, hydrophilic head. Phospholipid structure bilayers that acts as natural layers.

Phosphorylated Intermediate

A particle (frequently a reactant) with a phosphate group covalently bound to it, making it more responsive (less steady) than the unphosphorylated atom.

Photic Zone

The restricted top layer of a sea or lake, where light infiltrates adequately for photosynthesis to happen.

Photoautotroph

An organism that utilizes light energy to synthesise their own organic carbon through process of photosynthesis.

Photoheterotroph

An organism that utilisations light to produce ATP however don't use carbon di oxide as only carbon source for their growth and development.

Photoperiodism

A physiological reaction to photoperiod, the general lengths of night and day. An example of photoperiodism is flowering in plants.

Photophosphorylation

The method involved producing ATP from ADP and phosphate through chemiosmosis, utilising a proton-thought process force created across the thylakoid layer of the chloroplast or the film of specific prokaryotes during the light responses of photosynthesis.

Photorespiration

A metabolic pathway that consumes oxygen and ATP, discharges carbon dioxide and diminishes photosynthetic results. Photorespiration for the most part happens on hot, dry, brilliant days, when stomata close and the O2/CO2 proportion in the leaf increments, inclining toward the limiting of O2 as opposed to CO2 by rubisco.

Photosynthesis

The transformation of light energy to synthetic energy that is put away in sugars or other natural mixtures; happens in plants, green growth, and certain prokaryotes.

Photosystem

A light-catching unit situated in the thylakoid layer of the chloroplast or in the film of certain prokaryotes, comprising a responsible community complex encompassed by various light-collecting buildings. There are two kinds of photosystems, I and II; they assimilate light best at various frequencies.

Photosystem I (PS I)

A light-catching unit in a chloroplast's thylakoid layer or in the film of certain prokaryotes; it has two particles of P700 chlorophyll at its response community.

Photosystem II (PS II)

One of two light-catching units in a chloroplast's thylakoid layer or in the film of certain prokaryotes; it has two particles of P680 chlorophyll at its response community.

Phragmoplast

An arrangement of cytoskeletal components and Golgi-determined vesicles that structures across the midline of a partitioning plant cell during cytokinesis.

Phylocode

The International Code of Phylogenetic Nomenclature governs the nomenclature of organisms.Only groups that incorporate a typical ancestors and related family members are categorize in one taxa.

Phylogenetic Species Concept

Smallest group of individuals that share a sam ancestor, forming one clade.

Phylogenetic Tree

An extended outline or diagrams that depicts evolutionary history of different organisms of different genera and species.

Physical Map

A genetic map wherein the actual distances between genes or other hereditary markers are depicted, prepared based on the recombination frequency between different genes and markers.

Phytochrome

A type of light receptor in plants that for the most part assimilates red light and controls many plant reactions, like seed germination and shade aversion.

Phytoremediation

A method that aims to restore contaminated areas by exploiting a plant 's ability to extricate heavy metals and different contaminations from the soil and to accumulate them in effortlessly collected parts of the plant.

Pilus

In bacteria and other archea, a hair likestructure present on the surface helps in bacterial conjugation; otherwise called a sex pilus or formation pilus.

Pineal Gland

A little organ on the dorsal surface of the vertebrate forebrain that secretes the chemical melatonin.

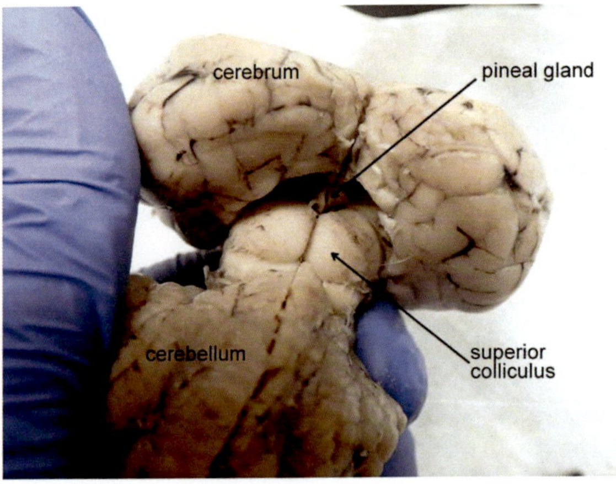

Figure 59. Brain Showing The Colliculi And Pineal Gland, Posterior View.

Source : Image By Flickr

PIR

Protein information resource, an annotated protein database similar to SWISS-PROT. PIR is also the name of a sequence format that is similar to FASTA

Pith

Ground tissue that is interior to the vascular tissue in a stem; in numerous monocot roots, parenchyma cells that structure the focal centre of the vascular chamber.

Pituitary Gland

An endocrine organ at the foundation of the nerve centre; comprises a back projection, which stores and deliveries two chemicals created by the nerve centre, and a foremost flap, which delivers and secretes a large number that controls different body functions.

Placenta

It is a structure in the pregnant uterus for supporting a viviparous hatchling with the mother's blood supply; developed from the uterine covering and undeveloped layers.

Plasma Cell

Plasma cell or effector B cells are white blood cells develops from B lymphocytes in bone marrow. Plasma cells developes from antigen-stimulated B cells. They are immunoagents helps in fighting infections/ diseases.

Plasma Membrane

The membrane found in each cell that separate the oute environment of the cell from inner environment, controlling the cell's synthetic organisation.

Plasmid

A small, circular, double stranded DNA molecule present in bacterial cells other than those of a bacterial chromosome; in DNA cloning, utilised as vectors transferring up to around 10,000 base pair (10 kb) of DNA. Plasmids are additionally found in certain eukaryotes, like yeasts.

Plasmodesma

An open channel through the cell wall associates the cytoplasm of neighbouring plant cells, permitting water, little solutes, and a few bigger particles to pass between the cells.

Plasmodial Slime Mold

A kind of protist that has amoeboid cells, whipped cells, and plasmodial taking care of stage in its life cycle.

Plasmodium

Obligate parasite of insects humans and other vertebrtes. It is a single cell eukarotic organism that grows on other organisms for their carbon requirement.

Plasmogamy

In organisms, the combination of the cytoplasm of cells from two individuals; happens as one cycle of sexual reproduction, followed later by karyogamy.

Plasmolysis

A peculiarity in walled cells wherein the cytoplasm shrink and the plasma layer pull from the cell wall; happens when the cell loses water to a hypertonic solution.

Plastid

One of a group of firmly related organelles that incorporates chloroplasts, chromoplasts, and amyloplasts. Plastids are found in cells of photosynthetic eukaryotes.

Plate Tectonics

The hypothesis is that the landmasses are essential for extraordinary plates of Earth's outside layer that float on the hot, fundamental piece of the mantle. Developments in the mantle make the landmasses move gradually over the long haul.

Platelet

Platelets are blood cells developed inspecific bone marrow cell. Platelets circle in the blood and are important for blood coagulation.

Polar Covalent Bond

A covalent bond between moleculaes that differs in electronegativity. The common electrons are pulled nearer to the more electronegative particle, making it marginally negative and the other molecule somewhat certain.

Polar Molecule

A particle (like water) with lopsided dissemination of charges in various area of the atom.

Polarity

An absence of evenness; underlying difference in the furthest poles of an organic molecule or structure, for example, the root end and shoot end of a plant.

Pollen Grain

In seed plants, a structure comprising of the male gametophyte..

Figure 60. Image Showing Pollen Grains.

Source : Image By Wikimedia Commons

Pollen Tube

A cylinderical tube like structure helps in the transfer of sperm to the ovule.

Pollination

The process of transferringpollen to the female reproductive part of a seed plant containing the ovules, leading to fertilization.

Polygamous

Alluding to a type of relationship wherein a person of one sex mates with a few of the other.

Polymer

A long molecule comprising of numerous comparatives or indistinguishable monomers connected together by covalent bonds.

Polymerase Chain Reaction (PCR)

A procedure for amplifying DNA *in vitro* by reacting it with explicit mastermix, a DNA polymerase enzyme, and nucleotide base pairs.

Polynucleotide

A polymer comprising numerous nucleotide monomers in a chain. The nucleotides can be those of DNA or RNA.

Polyploidy

A chromosomal modification wherein the organism has in excess of two complete chromosome sets. It is the aftereffect of a irregularity of cell division.

Polytomy

In a phylogenetic tree, a branch point from which multiple relative taxa arise. A polytomy demonstrates that the evolutionary connections between the relative taxa are not yet clear.

Pons

A part of the central nervous system connects the cerebellum with the medula that takes an interest in specific programmed, homeostatic function.

Population

A group of individuals of the very species that live in a similar region and interbreed, delivering healthy progenies.

Population Dynamics

The investigation of how complex collaborations among biotic and abiotic factors impact varieties in populace size.

Population Ecology

The investigation of populaces corresponding to their current environment, remembering ecological impacts for populace thickness and dissemination, age, structure, and varieties in populace size.

Positive Feedback

A type of guideline wherein a finished result of an interaction speeds up that cycle; in physiology, a control instrument in which an adjustment of a variable triggers a reaction that builds up or intensifies the change.

Posterior Pituitary

Augmentation of the nerve centre is made out of sensory tissue that secretes oxytocin and antidiuretic chemicals made in the nerve centre; an impermanent stockpiling site for these chemicals.

Postzygotic Barrier

A regenerative obstruction that forestalls half and half zygotes delivered by two distinct species from forming into reasonable, prolific grown-ups.

Potential Energy

The energy that matter has because of its area or spatial course of action (structure).

Pressure Potential (ΨP)

A part of water expected comprises of the actual tension on an answer, which can be positive, zero, or negative.

Prezygotic Barrier

A regenerative barrier that blocks mating between species or interfere in interspecific mating efforts.

Primary Cell Wall

In plants, a moderately slender and adaptable layer surround the plasma membrane of a young cell.

Primary Electron Acceptor

In the thylakoid layer of a chloroplast or in the membrane of certain prokaryotes, a particular molecule that shares the response group complex with a couple of chlorophyll atoms accepts an electron from them.

Primary Immune Response

The underlying versatile insusceptible reaction to an antigen shows up after a period of around 10 to 17 days.

Primary Producer

An autotroph is typically a photosynthetic organism. By and large, autotrophs make up the top level of a biological system that eventually upholds any remaining levels.

Primary Production

How much light energy is switched over completely to synthetic energy (natural mixtures) by the autotrophs in a biological system during a given time frame?

Primary Succession

A sort of environmental progression that happens in a space where there were initially no organisms present and where the soil has not yet been formulated.

Primary Transcript

An underlying RNA transcript from any gene; additionally, called pre-mRNA when translated from a protein-coding sequence.

Primary Visual Cortex

The objective is the occipital projection of the frontal cortex for the vast majority of the axons from the horizontal geniculate cores.

Primase

A compound that joins RNA nucleotides to make a preliminary during DNA replication, utilising the parental DNA strand as a template.

Primer

A short stretch of RNA with a free 3 end, limited by corresponding base matching to the template strand and stretched with DNA nucleotides during DNA replication.

Primitive Streak

A thickening along with the future foremost back pivot on the outer layer of an early avian or mammalian undeveloped organism is brought about by a stacking up of cells as they assemble at the midline prior to moving into the undeveloped organism.

Prion

Prions are abnormal proteins that are transmissible and have ability to transmit the abnormal protein in place of normal variant of protein leading to fatal neurodegenerative disease.

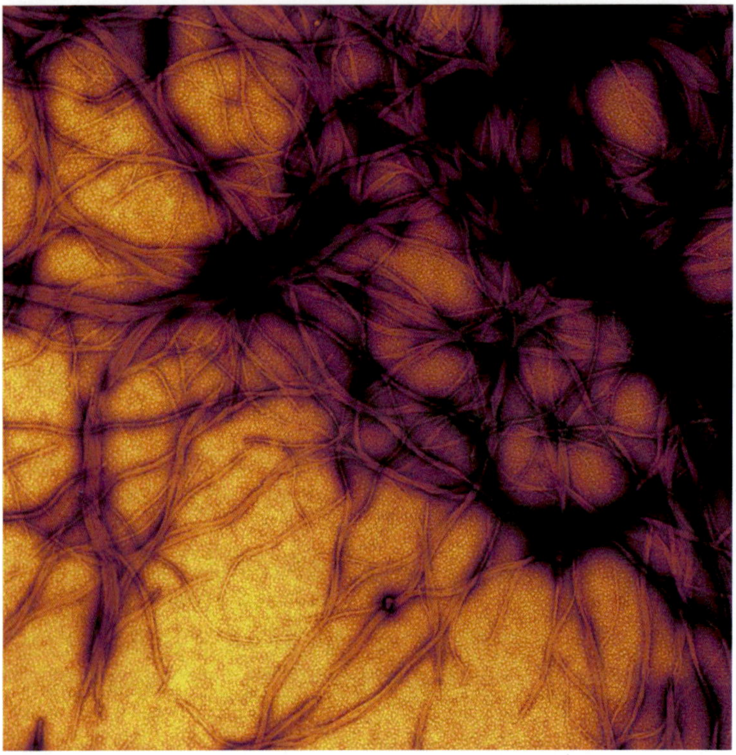

Figure 62. Prion Protein Expressed In E. Coli.

Source : Image By Flickr

Problem Solving

The mental movement of formulating a technique continues starting with one state and then onto the next notwithstanding genuine or evident obstructions.

Producer

An organism that produces natural mixtures from CO2 by utilizing light energy (in photosynthesis) or by oxidising inorganic synthetic substances (in chemosynthetic activity performed by certain prokaryotes).

Production Efficiency

The level of energy put away in acclimatised food that isn't utilised for breath or dispensed with as waste.

Progesterone

A steroid chemical that helps the uterus for maintaining pregnancy in evolved animals.

Profile

A sequence profile is usually derived from multiple alignments of sequences with a known relationship, and consists of a table of position - specific scores and gap penalties. Each position in a profile contains scores for all possible amino acids, as well as one penalty score for opening and one for continuing a gap at the position specified.

Prokaryotic Cell

A type of cell that lacks nucleus and membrane enclosed organelles. Organisms with prokaryotic cells (microorganisms and archaea) are called prokaryotes.

Prolactin

A chemical delivered and discharged by the foremost pituitary with an extraordinary different functions in various vertebrate species. In well-evolved animals, it induces the development of and milk production by the mammary glands.

Prometaphase

The second phase of mitosis, wherein the nuclear membrane breaks and the shaft microtubules connect to the kinetochores of the chromosomes.

Promiscuous

Alluding to a kind of relationship wherein mating happens casually or randomly.

Promoter

A particular nucleotide sequence in the DNA of a gene that binds with RNA polymerase, situating it to begin translating RNA at the suitable point.

Prophase

It is the first phase of mitosis, in which the chromatin consolidates into discrete chromosomes noticeable with a light microscope only, the mitotic shaft starts to develope, and the nucleolus vanishes however the nucleus stays in salvageable shape.

Prostaglandin

It is a chemical derived from fatty acid arachidonic acid produced in reponse to cell membrane injury by all the cells of the body.

Proteasome

A protein complex that regulates degradation of harmful unwanted proteins by the process of proteolysis.

Protein

An organically derived molecule comprising of at least one polypeptide collapsed and configured into a specific three-dimensional structure.

Protein Data Bank (PDB)

Database of every protein for which the 3D structure is known, it also contains a few non-protein structures.

Protein ID (in GenBank)

An identification number assigned to the amino acid sequence data included within a sequence record. This sequence identifier uses the accession version format. Each protein ID is made up of three letters, followed by five digits, a period, and a version number. For example, in sequence record M12345, the Protein ID for the sequence translation could be AAA35650.

Protein Kinase

A chemical that moves phosphate bunches from ATP to a protein, along these lines phosphorylating the protein.

Protein Mass Fingerprinting (PMF)

Identification of proteins by matching the masses of peptides contained in the protein with those of known proteins in a database.

Protein Microarrays

Proteomic methods similar to DNA microarrays in size and scale' proteins are spotted onto glass and are used to determine protein interaction or to identify and quantify molecules found in various solutions.

Protein Profile

A tool for visualizing a particular property (hydrophathy, charge, etc.) along a particular sequence by using a sliding window technique.

Protein Phosphatase

A chemical that eliminates phosphate bunches from (dephosphorylates) proteins, frequently working to turn around the impact of a protein kinase.

Proteoglycan

A macromolecule comprising of a little core protein with one or more glycosaminoglycan chains is found in the extracellular matrix of animal cells. A proteoglycan may comprise up to 95% starch.

Protist

A casual term applied to any eukaryote that isn't a plant, animal, or organism. Most protists are unicellular; however, some are provincial or multicellular.

Proton

A subatomic molecule with a single positive electrical charge, with a mass of around 1.7 * 10-24 g, is found in the core of a molecule.

Proton Pump

A functioning vehicle protein in a cell wall that utilises ATP to move hydrogen particles out of a cell against their fixation inclination, producing a membrane potential simultaneously.

Protonephridium

An excretory tube likestructure devoid of interior openings. It is present in rotifers, Platyhelminthes and few chordates.

Proton-Motive Force

The potential energy also reffered as a proton electrochemical gradient, produced by the removal of hydrogen particles (H+) across a natural membrane during chemiosmosis.

Protoplast Fusion

The fusion of two protoplasts from different plant species to generate interspecific hybrid by plant tissue culture.

Protostome Advancement

In animals, a formative mode recognised by the advancement of the mouth from the blastopore. In such organism formation of mouth occur before anus during embryonic developement

Proximal Tubule

In the vertebrate kidney, the part of a nephron promptly downstream from Bowman's capsule transfers and filter filtrate.

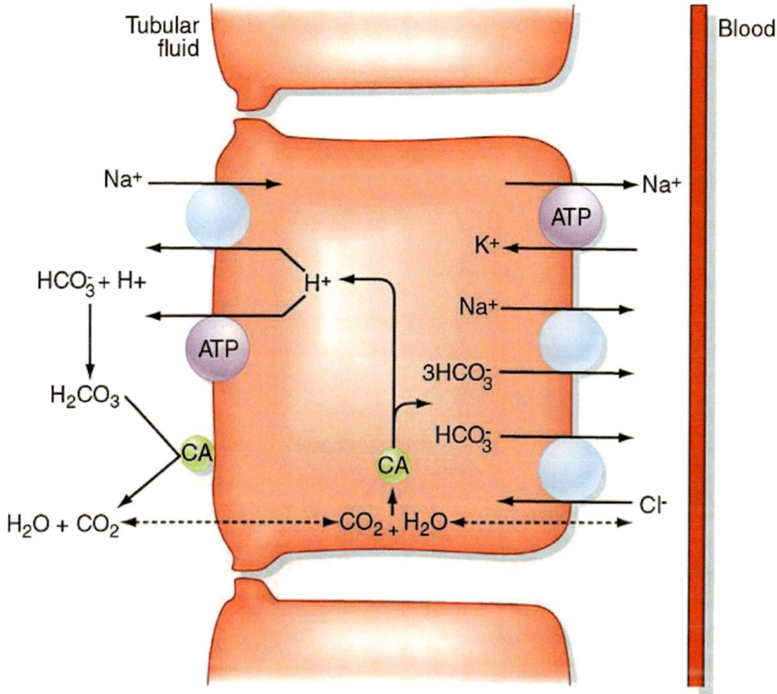

Figure 63. Proximal Convoluted Tubule.

Source: Image by Wikimedia Commons

Pseudocoelomate

An animal whose body does not derived from mesoderm.

Pseudogene

A DNA fragment basically the same as a pure gene however it doesn't yield a function; a DNA section that previously worked as a gene yet has become inactivated in specific animal categories as a result of transformation.

Pterophyte

A nickname for an individual from the phylum Pterophyta, which incorporates algae, horsetails, and whisk plants and their family members.

Pubmed

An extensive database of biomedical literature hosted by NCBI that is searchable.

Punnett Square

It is a diagram to depict the genotype of progenies of a cross between two homozygous parents.. It is named after its discoverer Reginald C. Punnett.

Pupil

The opening in the iris concedes light into the inside of the vertebrate eye. Muscles in the iris control their size.

Purine

One of two kind of nitrogenous bases found in nucleotides, described by a six-membered ring intertwined with a five-membered ring. Adenine (A) and guanine (G) are purines.

Pyrimidine

One of two kinds of nitrogenous bases found in nucleotides, portrayed by a six-membered ring. Cytosine (C), thymine (T), and uracil (U) are pyrimidines.

P-Value

Probability associated with a statistical test of the difference between populations. Populations are considered significantly different if the associated p-value is small (typically 0.1 or smaller).

Q

Quantitative Character

A heritable character controlled by many genes or alleles with small and cumulative effects.

Quaternary Structure

It is a structure of proteins that consists two or more polypeptide chains. The arrangement of several tertiary proteins forms quaternary structure.

Query (Sequence)

A DNA, RNA of protein sequence used to search a sequence database in order to identify close or remote family members (homologs) of known function, or sequences with similar active sites or regions (analogs), from whom the function of the query may be deduced.

R

Radial Cleavage

A sort of early-stage improvement in deuterostomes in which the planes of cell division that change the zygote into a bundle of cells are either equal or opposite to the upward pivot of the undeveloped organism, along these lines adjusting levels of cells one over the other.

Radial Symmetry

Evenness in which the body is formed like a pie or barrel (coming up short on a left side and a right side) and can be partitioned into reflecting imaged parts by any plane through its focal pivot.

Radioactive Isotope

An isotope (a nuclear type of a synthetic component) that is temperamental; the core decays precipitously, radiating distinguishable molecules and energy.

Radiolarian

A protist, typically marine, with a shell commonly made of silica and pseudopodia that transmit from the focal body.

Radiometric Dating

A technique for deciding the outright period of rocks and fossils, in light of the half-existence of radioactive isotopes.

Ras Gene

A gene that codes for proteins involved in cell signalling pathway.

Ray-Finned Fish

An individual from the class Actinopterygii, amphibian osteichthyans with fins which are supported by bony spines, includes fish, bass, and herring.

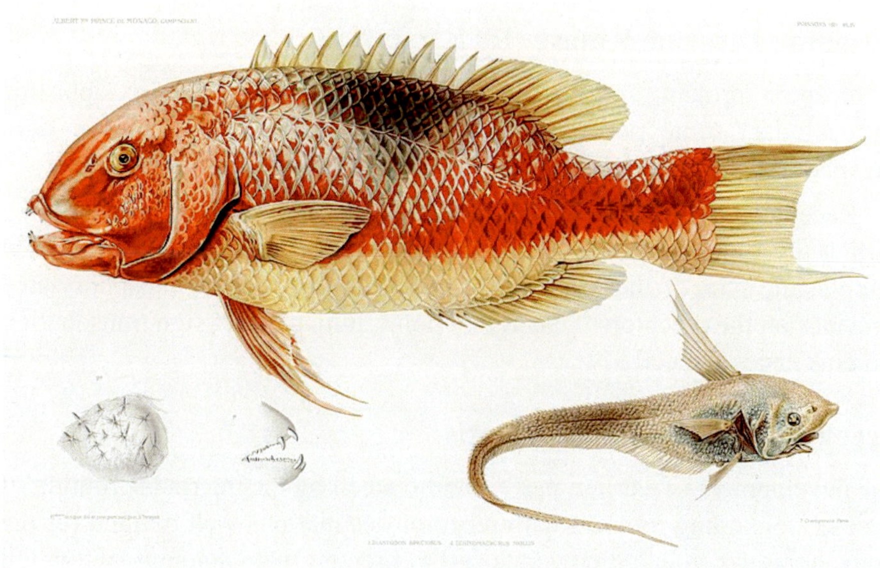

Figure 64. Hogfish And A Ray Finned Fish.

Source : Image By Wikimedia Commons

Response Center Complex

A complex of proteins related with extraordinary sets of chlorophyll molecules and an essential electron acceptor. Found halfway in a photosystem, this complex trigger the light responses of photosynthesis. Induced by light energy, the sets of chlorophylls give an electron to the essential electron acceptor, which passes an electron to an electron transport chain.

Reading Frame

On an mRNA, the triplet of ribonucleotides is utilised by the translation apparatus during polypeptide synthesis.

Receptor Potential

An underlying reaction of a receptor cell to an improvement, comprising of an adjustment of voltage across the receptor layer corresponding to the increased potential.

Receptor Tyrosine Kinase (RTK)

A receptor protein spreads over the plasma layer, the cytoplasmic (intracellular) some portion of which can catalyse the exchange of a phosphate group from ATP to tyrosine on another protein.

Receptor tyrosine kinases frequently answer the limiting of a flagging particle by dimerising and afterwards phosphorylating a tyrosine on the cytoplasmic part of the other receptor in the dimer. The phosphorylated tyrosines on the receptors then, at that point, initiate other sign transduction proteins inside the cell.

Receptor-Mediated Endocytosis

The development of explicit particles into a cell by the internal sprouting of vesicles containing proteins with receptor destinations well defined for the atoms being taken in; empowers a cell to procure mass amounts of explicit substances.

Reciprocal Altruism

Philanthropic way of behaving between inconsequential people, by which the charitable individual advantages later on when the recipient responds.

Recombinant Chromosome

A chromosome made while crossing over joins DNA from two parents into a single chromosome.

Recombinant Type (Recombinant)

A progeny whose chromosome differs from that of the pure reproducing parents; additionally, alludes to the actual chromosome.

Recursion

An algorithmic procedure whereby an algorithm calls on itself to perform a calculation until the result exceeds a threshold, in which case the algorithm exits. Recursion is a powerful procedure with which to process data and is computationally quite efficient.

Red Alga

A photosynthetic protist, named for its tone, results from a red colour that covers the green chlorophyll. Most red green growth is multicellular and marine.

Figure 65. Red Algae.

Source : Image By Flickr

Redox Reaction

A substance response includes the total or fractional exchange of at least one electron starting with one reactant and then onto the next; short for decreased oxidation response.

Refractory Period

The brief time frame following an activity potential in which the neuron can't answer another improvement is attributable to the inactivation of voltage-gated sodium channels.

Regulatory Gene

A gene that codes for a protein, like a repressor, controls the expression of one or more gene or group of genes.

Reinforcement

In transformative science, an interaction in which a cycle in which regular determination reinforces prezygotic obstructions to reproduction, along these lines diminishing the possibilities of half and half arrangement. Such an interaction is probably going to happen provided that half and half progenies are less fit than individuals from the parent species.

Relational Database

A database that follows E. F. Codd ' s 11 rules, a series of mathematical and logical steps for the organization and systemization of data into a software system that allows easy retrieval, updating, and expansion.

Relational Database Management Systems (RDBMS)

A software system that includes database architecture, query language, and data loading and updating tools and other ancillary software that together allow the creation of a relational database application. An RDBMS stores data in a database consisting of one or more tables of rows and columns. The rows correspond to a record (tuple); the columns correspond to attributes (fields) in the record. In an RDBMS, a view, defined as a subset of the database that is the result of the evaluation of a query, is a table. RDBMSs use Structured Query Language (SQL) for data definition, data management, and data access and retrieval. Relational and object - relational databases are used extensively in bioinformatics to store sequences and other biological data.

Relative Fitness

The fitness of a single individual in population is more genetically comparative with the fitness of others in the populace.

Renal Pelvis

The funnel shaped area of through which urine flows from kidey to the ureter.

Renin-Angiotensin-Aldosterone System (Raas)

A chemical regulating pathway that controls circulatory blood pressure and blood volume.

Repeats (Repeat Sequences)

Repeat sequences and approximate repeats occur throughout the DNA of higher organisms (mammals). For example, Alu sequences of about 300 characters in length appear hundreds of thousands of times in human DNA, with about 87% homology to a consensus Alu string. Some short substrings, such as TATA - boxes, poly - A, and (TG) * , also appear more often than would be expected by chance. Repeat sequences may also occur within genes, as mutations or alterations to those genes. Repetitive sequences, especially mobile elements, have many applications in genetic research. DNA transposons and retroposons are used routinely for insertional mutagenesis, gene mapping, gene tagging, and gene transfer in several model systems.

Repetitive DNA

Nucleotide arrangements, typically noncoding, are available in many duplicates in a eukaryotic genome. The rehashed units might be short and organized tandemly (in series) or long and scattered in the genome.

Replication Fork

A Y-molded district on a duplicating DNA particle where the parental strands are being loosened up and new strands are being orchestrated.

Repressor

A protein that represses gene function In prokaryotes, repressors bind to the DNA in or close to the promotor. In eukaryotes, repressors might bind to control components inside enhancers, to activators, or to different proteins that block activators from restricting DNA.

Reproductive Isolation

The presence of natural variables (obstructions) that hinder individuals from two species from delivering healthy, prolific progenies.

Resource Partitioning

The division of ecological assets by existing together species to such an extent that the speciality of every species differ by at least one critical variable from the specialities of every single coinciding specie.

Response

(1) In cell correspondence, the adjustment of a particular cell movement is achieved by a transduced signal from outside the cell.

(2) In the criticism guidelines, a physiological movement is set off by an adjustment of a variable.

Resting Potential

The wall expected attributes of a nonconducting sensitive cell, within the cell are more negative than on the outside.

Restriction Enzyme

An endonuclease (sort of chemical) that perceives and cuts DNA frangments unfamiliar to a bacterium (like phage genomes). The compound cuts at explicit nucleotide arrangements.

Restriction Fragment Length Polymorphism (RFLP)

It is technique that uses variation in nucleotide sequences generated from cutting the total DNA of an organism using specific restriction endonuclease enzyme.

Restriction Map

A physical map or depiction of a gene (or genome) derived by ordering overlapping restriction fragments produced by digestion of the DNA with a number of restriction enzymes.

Reticular Formation

A diffuse organisation of neurons in the centre of the brainstem that channels data venturing out to the cerebral cortex.

Retina

The deepest layer of the vertebrate eye, containing photoreceptor cells (poles and cones) and neurons; communicates pictures framed by the focal point to the cerebrum by means of the optic nerve.

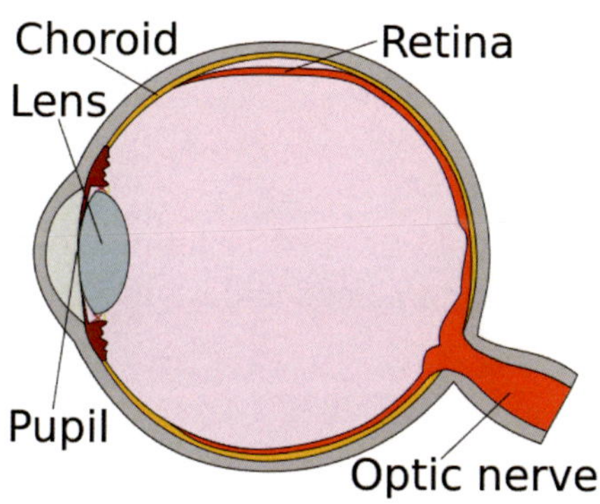

Figure 66. Cross Section Of A Human Eye, Showing A Detached Retina At Top.

Source : Image By Wikimedia Commons

Retrotransposon

A transposable component that moves inside a genome through an RNA by converting RNA into DNA through reverse transcription process.

Retrovirus

An RNA infection that duplicates by translating its RNA into DNA and afterwards embedding the DNA into a cell chromosome; is a significant class of malignant growth causing infections.

Reverse Transcriptase

A compound encoded by certain infections (retroviruses) that involves RNA as a template for DNA synthesis.

Reverse Transcriptase–Polymerase Chain Reaction (RT-PCR)

A procedure for deciding function of a specific gene. It utilises reverse transcriptase and DNA polymerase enzymes to orchestrate cDNA from all the mRNA in an respected sample and afterwards subjects the cDNA to PCR amplification involving preliminaries explicit for the the duplication of gene of interest.

Rhizaria

One of five supergroups of eukaryotes proposed in ongoing speculation of the transformative history of eukaryotes; is a morphologically different protist clade that is characterised by DNA similitudes.

Rhizobacterium

A soil bacterium whose populace size is tremendously increased in the rhizosphere, the soil area near a plant's underlying roots.

Rhizoid

A long, cylindrical single cell or fibre of cells that secures bryophytes to the ground. Not at all like roots, rhizoids are not made out of tissues, need particular directing cells, and don't assume an essential part in water and mineral assimilation.

Rhodopsin

A visual shade comprising retinal and opsin. After engrossing light, the retinal changes shape and separates from the opsin.

Rhythm Method

A type of contraception that depends on abstaining from sex when origination is probably going to happen; is additionally called regular family arranging.

Ribonucleic Acid (RNA)

A kind of nucleic acid comprising of a polynucleotide having nucleotide monomers with a ribose sugar and the nitrogenous bases adenine (A), cytosine (C), guanine (G), and uracil (U); typically single-stranded; function in protein synthesis, gene regulation, and as the genome of some infectious agents.

Ribosomal RNA (rRNA)

RNA molecule that, along with proteins, makeup ribosomes; the most abundant type of RNA.

Ribosome

A complex of rRNA and protein molecule that helps as a site of protein synthesis in the cytoplasm; comprises an big and a small subunit. In eukaryotic cells, every subunit is collected in the nucleolus.

Ribozyme

An RNA particle that functions as a catalyst, for example, an intron that catalyses its own evacuation during RNA removal.

RNA Interference (RNAi)

A procedure used to silent the function of chosen gene. RNAi utilises synthetic double stranded abandoned RNA molecule that match the sequence of a specific gene to silent the same.

RNA Polymerase

A chemical that joins ribonucleotides into a developing RNA chain during translation, in light of corresponding restricting to nucleotides on a DNA template strand.

RNA Processing

Adjustment of RNA essential function, including removal of introns, joining of exons, and modification of the 5 and 3 end.

RNA Splicing

After the synthesis of a eukaryotic essential RNA strand, the removal of segments of the strand (introns) that won't be code for the mRNA and the association of the remaining parts (exons).

Root

An organ in vascular plants that secures the plant and empowers it to assimilate water and minerals from the soil.

Figure 67. Roots Of A Plant.

Source : Image By Pixabay

Root Hair

A small augmentation of a root epidermal cell, developing simply behind the root tip and expanding surface region for assimilation of water and minerals.

Root Pressure

Pressure applied in the underlying root of plants is the aftereffect of assimilation, causing exudation from cut stems and guttation of water from leaves.

Root System

The entirety of a plant's underlying root, anchor it in the soil, assimilate and transport minerals and water, and store food.

Rooted

Portraying a phylogenetic tree that contains a branch point (frequently, the one farthest to one side) addressing the latest normal predecessor of all taxa in the tree.

Round Window

In the mammalian ear, the resource where vibrations of the stapes make a voyaging series of tension waves in the liquid of the cochlea.

Rubisco

Ribulose bisphosphate (RuBP) carboxylase, is the chemical that catalyzes the initial step of the Calvin cycle (the expansion of $CO2$ to RuBP).

Ruminant

An animal, like a cow or a sheep, with different stomach compartments particular for a herbivorous eating routine.

S

Saccule

In the vertebrate ear, a chamber in the vestibule behind the oval window takes an interest in the feeling of equilibrium.

Salicylic Acid

A flagging particle in plants that might be to some degree liable for initiating foundational obtained protection from microorganisms.

Salivary Gland

An organ related to oral depression secretes substances that grease up food and start the course of artificial assimilation.

Saltatory Conduction

Quick transmission of a nerve motivation along an axon, coming about because of the activity expected bouncing starting with one hub of Ranvier then onto the next, avoiding the myelin-sheathed districts of the layer.

SAM File

A text format for storing sequence data in a series of tab delimited ASCII columns.

Sarcoplasmic Reticulum (SR)

A particular endoplasmic reticulum controls the calcium fixation in the cytosol of muscle cells.

Saturated Fatty Acid

An unsaturated fat wherein all carbons in the hydrocarbon tail are associated by single securities, along these lines augmenting the number of hydrogen molecules that are connected to the carbon skeleton.

Savanna

A tropical prairie biome with dissipated individual trees and enormous herbivores and kept up with intermittent flames and dry season.

Scaffolding Protein

A kind of huge transfer protein to which a few other hand-off proteins are at the same time connected, expanding the effectiveness of sign transduction.

Scanning Electron Microscope (SEM)

A microscope that utilises an electron pillar to examine the outer layer of a sample, covered with metal molecules, to concentrate on subtleties of its geography.

Schizophrenia

An extreme mental unsettling influence is described by maniacal episodes in which patients have a contorted impression of the real world.

Schwann Cell

A type of glial cell that structures protecting myelin sheaths around the axons of neurons in the fringe sensory system.

Sclereid

A short, unpredictable sclerenchyma cell in nutshells and seed coats. Sclereids are dissipated all through the parenchyma of certain plants.

Sclerenchyma Cell

An inflexible, strong plant cell type is typically inadequate with regards to a protoplast and has thick auxiliary walls reinforced by lignin at development.

Scrotum

A pocket of skin outside the mid-region that houses the testicles; helps in keeping up with the testicles at the lower temperature expected for spermatogenesis.

Second Messenger

A little, nonprotein, water dissolvable molecule, like a calcium particle (Ca2) or cyclic AMP, that are intracellular signalling molecule that signals the cell surface receptors to effector protein.

Secondary Cell Wall

In plant cells, a solid and strong structure that are developed overtime around cell wall and plasma membrane give strength and mechanical support.

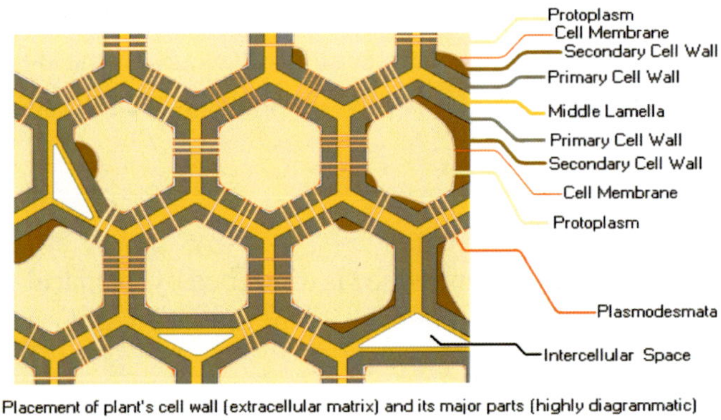

Figure 68. Plant Cell Wall.

Source : Image By Wikimedia Commons

Secondary Endosymbiosis

An interaction in eukaryotic development wherein a heterotrophic eukaryotic cell engulf a photosynthetic eukaryotic cell, which made due to an advantageous relationship inside the heterotrophic cell.

Secondary Growth

Secondary growth development is performed by horizontal meristems, thickening the roots and shoots of woody plants.

Secondary Immune Response

The versatile immune response is induced by second or repeated entry to a specific antigen. The secondary immune response is fast, and more efficient in eliminating antigens.

Secretion

(1) The release of substrates initiated by a cell.

(2) The release of wastes from the body liquid into the filtrate.

Secretory Phase

That part of the uterine (feminine cycle) when the endometrium keeps on thickening, turns out to be more vascularized and creates organs that discharge a liquid high in glycogen.

Seed

seed is matured ovule generated from the fertilization of male and female reproductive cells. Seed is protected by outer covering known as seed coat.

Seed Coat

An outer covering of a seed, developed from the outer layer of an ovule. In a flowering plant, the seed coat encloses and safeguards the undeveloped embryo and endosperm.

Seedless Vascular Plant

Seedless vascular plants are reproduces through spores and doesnot produce seeds. structure a paraphyletic group that incorporates the phyla Lycophyta (group of algae and their family members) and Pterophyta (greeneries and their family members).

Selective Permeability

A property of organic membranes that permits them to control the type of substances that pass through them.

Self-Incompatibility

The inability of pollen to fertilize the egg of same plant and some times the pollen from the plants of related family also not compatible for fertilization.

Semelparity

It is a condition in which an organism is able to reproduce and produce progeny only once in its life cycle .

Semen

A fluid that is discharged by the male reproductive organsand contains sperm that fertilizes with egg to produce zygote.

Semiconservative Model

A type of DNA replication in which the replicated double standed helix comprises one old strand recieved from the parent, and one recently made strand.

Semilunar Valve

A valve is situated at each exit of the heart, where the aorta leaves the left ventricle and the aspiratory supply route leaves the right ventricle.

Seminal Vesicle

An organ in males that secretes a semen fluid that greases up and sustains sperm.

Sensitivity

The sensitivity of bioinformatics similarity search algorithms centers around two areas: how well the method can detect biologically meaningful relationships between two related sequences in the presence of mutations and sequencing errors; and how the heuristic nature of the algorithm affects the probability that a matching sequence will not be detected. At the uscr' s discretion, the speed of most similarity search programs can be sacrificed in exchange for greater sensitivity — with an emphasis on detecting lower - scoring matches.

Sensory Adaptation

Reduced sensitivity to external stimulus over a period of time.

Sensory Neuron

A nerve cell that gets activated from the interior or outside environment and communicates signs to the focal sensory system.

Sensory Receptor

An organ, cell, or structure inside a cell that responses explicit improvements from an organism's outside or interior environment.

Sensory Transduction

The translation of sensory stimulous to neuronal activity.

Septum

A wall that devides the larger cavity into two smaller ones.

Serial Endosymbiosis

It provide explantion for the initiation of eukaryotes comprised of an arrangement of endosymbiotic occasions in which mitochondria, chloroplasts, and maybe other cell structures were developed from smaller prokaryotes.

Serotonin

It is a neurotransmitter that fuctions as mood stabilizer. It transmits messages between nerve cells in brain and human body.

Server

A combination of both computer hardware and software that supplies a service to requests submitted by client computers. The server processes the requests and then returns the results to the client. In short, it is a computer that makes services available on a network.

Sex-Linked Gene

A gene present on either sex chromosome. Most sex-connected genes are on the X chromosome for example the disorder hemophilia and Fabry disease are X linked diseases.

Sexual Reproduction

A type of reproduction wherein two parents mate to give rise to progenies that has alleles from both the parents through the gametes.

Sexual Selection

A type of mating wherein individual with specific acquired attributes is almost certain than others to get mates.

Shannon Diversity

A record of local area variety represented by H and addressed by the situation H= - (pA ln pA + PB ln PB + PC ln PC + . . .), where A, B, C . . . are species, p is the overall overflow of every species, and ln is the regular logarithm.

Shared Ancestral Character

In an individual, the trait is shared by all the organisms of specific clade, predecessor that isn't an individual from that clade.

Short Tandem Repeat (STR)

A sequence of DNA having tandemly repeated units of 2 to 5nucleotides. Differencess in STRs serve as hereditary markers in STR study, used to get hereditary profiles of an organism.

Short-Day Plant

A plant that flowers (for the most part in pre-fall, fall, or winter) just when the light period is shorter than the darker period.

Short-Term Memory

The ability to remember information, expectations, or objectives for a period and afterwards delete them on the off chance that they become insignificant.

Sickle-Cell Disease

An innately acquired human blooddisease wherein a single nucleotide changes in the - globin gene affecting haemoglobin , changing red blood cell shapesimilar to sickle hence the name is sickle cell anemia. .

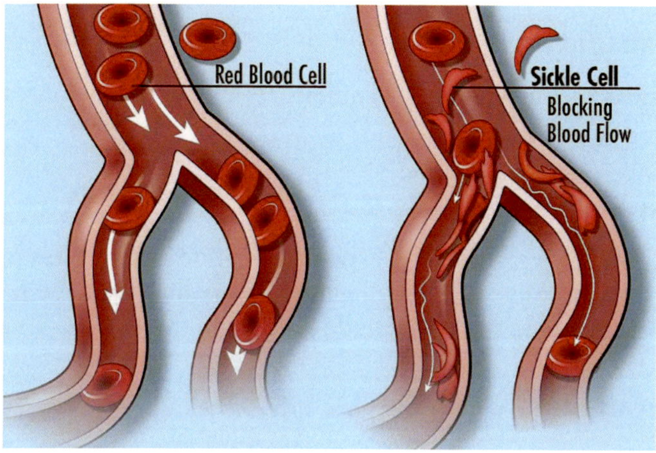

Figure 69. Normal Blood Cells (Left) And The Blood Cells In Sickle Cell Disease.

Source : Image By Wikimedia Commons

Sieve Plate

It is a end wall with poresthat interconnect with protoplasm of adajacent sieve tubes, that helps in transport phloem sap in plants.

Sieve-Tube Element

A living cell that transport the sugars and other natural compounds in the phloem of angiosperms; additionally, called a sieve tube element. Associated end to end forming sieve tubes divided by sieve plates.

Signal

In animals it refers to transmission of informationThe term is additionally used with regards to communication e in different types of organic molecules and in cell-to-cell communication in every single multicellular organism.

Signal Peptide

Signal pepetides are short sequence of about 20-40 amino acids at N terminal of primary protein sequencethat that carries information for protein secretion.

Signal Transduction

The association of a mechanical, synthetic, or electromagnetic response to a particular cell reaction.

Signal-Recognition Particle (SRP)

A protein RNA complex that target specific peptide as it reaches to ribosome and guides the ribosome to the endoplasmic reticulum (ER) by restricting it to a receptor protein on the ER.

Silent Mutation

A nucleotide-pair replacement significantly affects the organism; for instance, inside a gene, a transformation that results in a codon that codes for a similar amino acid.

Single Circulation

A circulatory system comprising of a single circuit, wherein blood pumps from heart to gills for cleaning for where it is circulated to rest of the body. This system is found in certain fish wherein blood passes through heart only once.

Single Nucleotide Polymorphism (SNP)

A single base-pair site in a genome where nucleotide difference is observed in about 1% of the populace.

Single-Strand Binding Protein

During DNA replication, this protein binds to unpaired DNA strands, stabilising and holding them apart while they serve as templates for the production of complementary DNA strands.

Sinoatrial (SA) Node

The pacemaker, which is an area in the right atrium of the heart that controls the rate and timing of all cardiac muscle cell contractions.

Sister Chromatids

Proteins near the centromere and, in some cases, along the arms attach two copies of a duplicated chromosome to one other. Two sister chromatids form one chromosome when they are linked together. During mitosis or meiosis II, the chromatids gradually split.

Sister Taxa

Groups of organisms that have a common ancestor and thus are close relatives.

Skeletal Muscle

A striated muscle that is responsible for the majority of the body's voluntary actions.

Sliding-Filament Model

Muscle contraction is thought to be based on the movement of thin (actin) filaments along thick (myosin) filaments, shortening the sarcomere, the fundamental unit of muscle architecture.

Slow Block To Polyspermy

The creation of the fertilisation envelope and other changes to an egg's surface that prevent multiple sperm from fusing with the egg. After fertilisation, the sluggish block begins about 1 minute later.

Small Interfering RNA (SiRNA)

From a long, linear, double-stranded RNA molecule, cellular machinery generates many tiny, single-stranded RNA molecules. The siRNA forms a complex with one or more proteins that can degrade or stop an mRNA with a complementary sequence from being translated. SiRNA can also inhibit transcription by increasing chromatin remodelling in rare instances.

Small Intestine

The alimentary canal's longest portion, so termed because of its diminutive diameter in comparison to the large intestine; the primary site of enzymatic breakdown of food macromolecules and nutrient absorption.

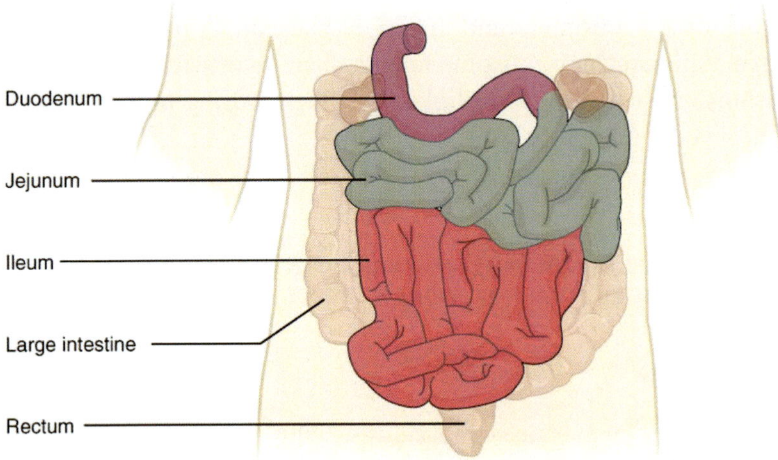

Figure 70. Small Intestine.

Source : Image By Wikimedia Commons

Smooth Muscle

Because of the homogeneous distribution of myosin filaments in the cells, this type of muscle lacks the striations of skeletal and cardiac muscle; it is responsible for involuntary body movements.

SMTP

Simple mail transfer protocol, defines the manner by which e-mail is transferred between computers on the Internet.

Sodium-Potassium Pump

A transport protein found in animal cells' plasma membranes that actively transfers sodium out of the cell and potassium in.

Solute Potential (ΨS)

It can be either zero or negative. A component of water potential that is proportional to the molarity of a solution and assesses the effect of solutes on the direction of water movement; also called osmotic potential.

Somite

In a vertebrate embryo, one of a sequence of mesoderm blocks that exist in pairs just lateral to the notochord.

Sorus

A sporangia cluster on a fern sporophyll. Sori can be arranged in a variety of patterns, such as parallel lines or dots, that help identify ferns.

Southern Blotting

The detection of specific nucleotide sequences in DNA samples using this approach. It entails the separation of DNA molecules on a gel and their blotting onto a membrane, followed by nucleic acid hybridization with a tagged probe.

Spatial Summation

The combined action of EPSPs or IPSPs produced nearly simultaneously by separate synapses determines the membrane potential of the postsynaptic cell, which is a phenomena of neuronal integration.

Species

A population or a set of populations whose members have the ability to interbreed in nature and create viable, fertile offspring, but who do not generate viable, fertile offspring with members of other populations.

Species-Area Curve

The pattern of biodiversity that illustrates that the wider a community's geographic area is, the more species it possesses.

Specific Heat

The quantity of heat required to raise or lower the temperature of 1 g of a substance by 1°C.

Spectrophotometer

The fractions of light of different wavelengths absorbed and transmitted by a pigment solution are measured with this equipment.

Sphincter

The size of an opening in the body, such as the passage between the oesophagus and the stomach, is controlled by a ring-like band of muscle fibres.

Spiral Cleavage

In protostomes, a type of embryonic development in which the planes of cell division that turn the zygote into a ball of cells are diagonal to the embryo's vertical axis. As a result, each tier's cells sit in the grooves between neighbouring tiers' cells.

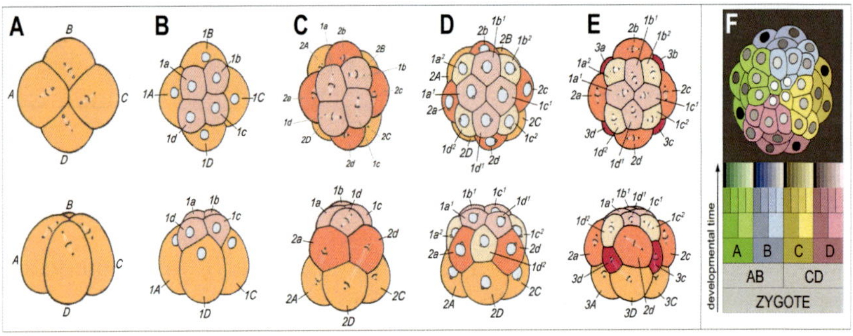

Figure 71. Spiral Cleavage in Gastropod *Trochus.*

Source : Image By Wikimedia Commons

Spliceosome

Splicing RNA occurs when a big complex of proteins and RNA molecules interacts with the ends of an RNA intron, releasing the intron and uniting the two neighbouring exons.

Spore

(1) A haploid cell formed in the sporophyte by meiosis throughout the life cycle of a plant or alga that undergoes generational alternation. A spore can divide into a multicellular haploid individual, the gametophyte, by mitosis

without merging with another cell. (2) In fungi, a haploid cell that creates a mycelium after germination, produced either sexually or asexually.

Sporophyte

The multicellular diploid form that develops from the union of gametes in organisms with alternation of generations (plants and some algae). By meiosis, the sporophyte generates haploid spores, which mature into gametophytes.

Sporopollenin

A long - lasting polymer that protects charophyte algal zygotes from drying out by forming the walls of plant spores.

Stabilizing Selection

Natural selection favours intermediate morphologies over extreme traits in terms of survival and reproduction.

Starch

In plants, a storage polysaccharide made completely of glucose monomers linked together by glycosidic connections.

Start Point

The nucleotide location on the promoter where RNA polymerase begins RNA synthesis in transcription.

Statocyst

Invertebrates have a form of mechanoreceptor called a statolith, which stimulates hair cells in proportion to gravity and keeps them in balance.

Statolith

(1) A specialised plastid in plants that contains dense starch granules and may play a role in gravity detection. (2) A dense particle found in sensory organs that act in equilibrium in invertebrates and settles in response to gravity.

Stem

A vascular plant organ that supports the leaves and reproductive systems through an alternating system of nodes and internodes.

Stem Cell

Any relatively unspecialized cell capable of producing one identical daughter cell and one more specialised daughter cell capable of further differentiation during a single division.

Steroid

A lipid with a carbon skeleton made up of four fused rings to which various chemical groups are connected.

Stoma

A microscopic opening in the epidermis of leaves and stems surrounded by guard cells that facilitates gas exchange between the environment and the plant's interior.

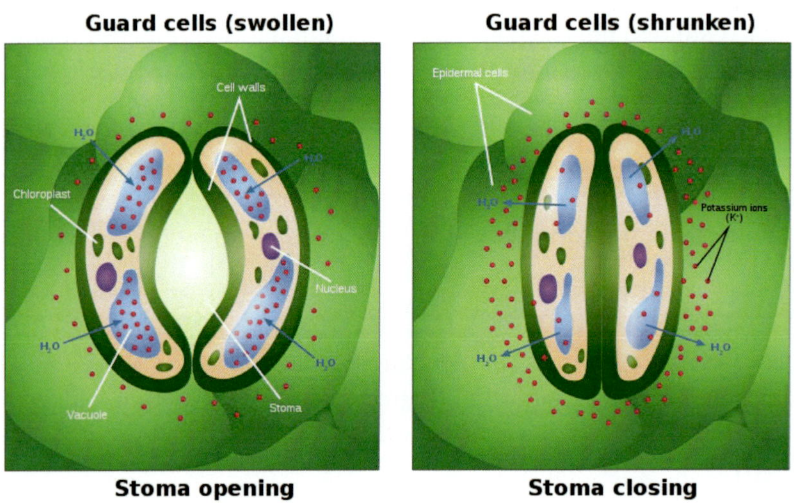

Figure 72. Opening And Closing of Stomata By Bean-Shaped Guard Cells.

Source : Image By Wikimedia Commons

Stramenopile

A protist having a flagellum that is "hairy" (covered with fine, hairlike projections) and a shorter, smooth flagellum.

Strigolactones

A type of plant hormone that slows shoot branching, increases the interaction of plant roots with mycorrhizal fungi, and triggers the germination of parasitic plant seeds.

Strobilus

In most gymnosperms and certain seedless vascular plants, the technical word for a cluster of sporophylls known as a cone.

Stroke

The death of nerve tissue in the brain, usually caused by a head artery rupture or obstruction.

Stroma

The dense fluid that surrounds the thylakoid membrane in the chloroplast and contains ribosomes and DNA, and is involved in the synthesis of organic molecules from carbon dioxide and water.

Stromatolite

The activities of prokaryotes bind thin sediment films together, resulting in layered rock.

Structural Isomer

One of a number of compounds with the same chemical formula but different atom covalent configurations.

Structure Prediction

Algorithms that predict the secondary, tertiary, and sometimes even quarternary structure of proteins from their sequences. Determining protein structure from a sequence has been dubbed " the second half of the genetic code " since it is the higher - level folded structure of a protein that governs how it functions as a gene product. As yet, most structure prediction methods have been only partially successful and typically work best for certain well - defined classes of proteins.

Substrate-Level Phosphorylation

In catabolism, the enzyme facilitated the synthesis of ATP by transferring a phosphate group from an intermediate substrate to ADP.

Sugar Sink

A plant organ that is either a net sugar consumer or a sugar storer. Sugar sinks supplied by phloem include growing roots, shoot tips, stems, and fruits.

Sugar Source

Sugar is produced by either photosynthesis or the breakdown of starch in this plant organ. Plants' principal sugar sources are mature leaves.

Surface Tension

A measurement of how difficult it is to stretch or break a liquid's surface. The hydrogen bonding of surface molecules gives water a high surface tension.

Survivorship Curve

One technique to portray age-specific mortality is to plot the number of members of a cohort who are still alive at each age.

Suspension Feeder

A sponge, clam, or baleen whale is an aquatic animal that eats by sifting microscopic creatures or food particles from the water.

Sustainable Development

Development that meets people's current demands without jeopardising future generations' ability to meet their own.

Symbiosis

An ecological interaction that exists between organisms of two different species that live in close proximity.

Sympathetic Division

The autonomic nervous system is divided into three divisions, each of which raises energy expenditure and prepares the body for action.

Synapse

The point at which a neuron communicates with another cell via a neurotransmitter or electrical connection across a tiny gap.

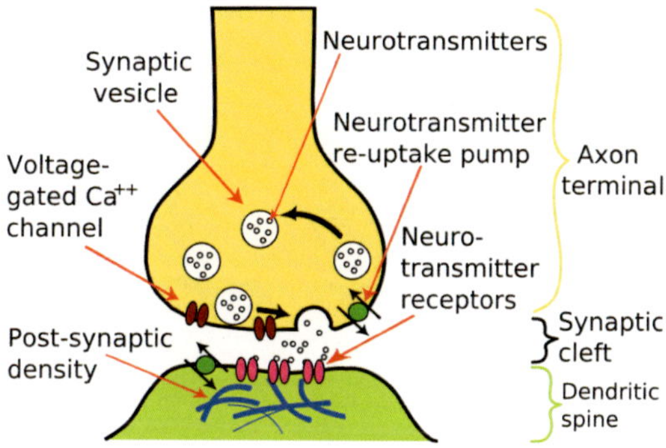

Figure 73. Synapse.

Source : Image By Wikimedia Commons

Synapsid

A single hole on each side of the skull distinguishes this amniote group. Mammalian synapsids include mammals.

Synapsis

During prophase I of meiosis, the pairing and physical interaction of duplicated homologous chromosomes.

Systematics

A branch of science concerned with identifying species and figuring out their evolutionary links.

Systemic Acquired Resistance

Infected plants' defensive reaction that aids in the protection of healthy tissue from pathogenic invasion.

Systemic Circuit

The branch of the circulatory system that delivers oxygenated blood to organs and tissues throughout the body and removes deoxygenated blood.

Systems Biology

An approach to biology that tries to simulate the dynamic behaviour of whole biological systems based on a study of the system's parts' interactions.An approach to biology that tries to simulate the dynamic behaviour of whole biological systems based on a study of the system's parts' interactions.

T

T Cells

The type of lymphocyte that matures in the thymus; it includes both effector cells for the cell-mediated immune response and helper cells for both branches of adaptive immunity.

Taproot

A primary vertical root that sprouts from an embryonic root and gives rise to lateral (branch) roots.

Figure 74. Tap Roots of a Wild Dandelion.

Source : Image By Flickr

Taste Bud

A group of modified epithelial cells on the tongue or in the mouth that serve as taste receptors in mammals.

Tay-Sachs Disease

A recessive allele for a malfunctioning enzyme causes a buildup of certain lipids in the brain in humans. Seizures, blindness, and deterioration of motor and mental ability typically appear a few months after birth, followed by death a few years later.

Technology

The use of scientific information for a specific purpose, usually in industry or commerce, but often in basic research.

Telomere

At the end of a eukaryotic chromosome's DNA molecule, there is tandemly repeating DNA. Telomeres safeguard the organism's genes from eroding over successive replication rounds.

Telophase

The fifth and final stage of mitosis, during which daughter nuclei are formed and cytokinesis is usually initiated.

Temperate Broadleaf Forest

A biome found in midlatitude locations with enough rainfall to allow the growth of large, broadleaf deciduous trees.

Template Strand

The DNA strand that serves as a pattern or template for organising the sequence of nucleotides in an RNA transcript via complementary base pairing.

Temporal Summation

A neural integration phenomenon in which the membrane potential of a postsynaptic cell in a chemical synapse is determined by the combined effect of EPSPs or IPSPs produced in quick succession.

Terminator

In bacteria, a sequence of nucleotides in DNA that signals RNA polymerase to release the freshly synthesised RNA molecule and detach from the DNA.

Territoriality

An animal's defence of a confined physical space against encroachment by other individuals, usually of its own species.

Tertiary Structure

The overall shape of a protein molecule as a result of amino acid side chain interactions such as hydrophobic interactions, ionic bonds, hydrogen bonds, and disulfide bridges.

Testcross

To determine the unknown genotype, an organism of unknown genotype is bred with a homozygous recessive individual. The offspring's phenotypic ratio exposes the unknown genotype.

Testosterone

The primary androgen in mammals; a steroid hormone essential for the development of the male reproductive system, spermatogenesis, and male secondary sex characteristics.

Tetanus

A skeletal muscle's maximal, prolonged contraction generated by an extremely high frequency of action potentials elicited by continuous stimulation.

Tetrapod

A vertebrate clade whose members have digitised limbs. Mammals, amphibians, birds, and other reptiles are all tetrapods.

Figure 75. Examples of Tetrapoda.

Source : Image By Wikimedia Commons

Thalamus

The vertebrate forebrain's integrating centre. Neurons with cell bodies in the thalamus relay neural input to specific sections of the cerebral cortex and govern which information is sent there.

Thallus

A plantlike seaweed body with a holdfast, stipe, and blades but no true roots, stems, or leaves.

Theory

An explanation that is more general than a hypothesis, creates new hypotheses, and is supported by a substantial body of evidence.

Thermodynamics

The investigation of energy transitions that take place in a collection of matter. See also the first and second laws of thermodynamics.

Thick Filament

A filament made up of staggered arrays of myosin molecules; a component of muscle fibre myofibrils.

Thigmo-Morphogenesis

A plant's response to persistent mechanical stimulation that results in enhanced ethylene synthesis. Thickening stems in response to severe winds is one example.

Thin Filament

A filament made up of two strands of actin coiled around two strands of regulatory protein; a component of myofibrils in muscle fibres.

Thylakoid

A membranous sac that is flattened inside a chloroplast. Thylakoids are frequently found in continuous stacks called grana; their membranes contain molecular "machinery" that converts light energy to chemical energy.

Thyroid Gland

On the ventral surface of the trachea, an endocrine gland that secretes two iodine-containing hormones, triiodothyronine (T3) and thyroxine (T4), as well as calcitonin.

Thyroxine (T4)

One of two iodine-containing hormones generated by the thyroid gland that aids in the regulation of metabolism, development, and maturation in vertebrates.

Ti Plasmid

A tumor-inducing bacterium's (the plant pathogen Agrobacterium) plasmid that integrates a piece of its DNA (T DNA) into a host plant's chromosome. The Ti plasmid is commonly employed as a vector in plant genetic engineering.

Tight Junction

An intercellular connection between animal cells that stops material from leaking through the space between cells.

Toll-Like Receptor (TLR)

A phagocytic white blood cell membrane receptor that identifies fragments of molecules shared by a group of pathogens.

Top-Down Model

Predation effects community organisation by controlling herbivore numbers, which in turn regulate plant or phytoplankton populations, which in turn control nutrient levels; also known as the trophic cascade model.

Topoisomerase

A protein that swivels, breaks, and rejoins DNA strands. Topoisomerase relieves tension in the double helix upstream of the replication fork during DNA replication.

Torsion

A developmental process in gastropods in which the visceral mass rotates up to 180°, allowing the anus and mantle cavity to be positioned above the animal's head.

Totipotent

A cell that can develop into all sections of the embryo and adult, as well as extraembryonic membranes in species that have them.

Trachea

The part of the respiratory tract that connects the larynx to the bronchi; sometimes known as the windpipe.

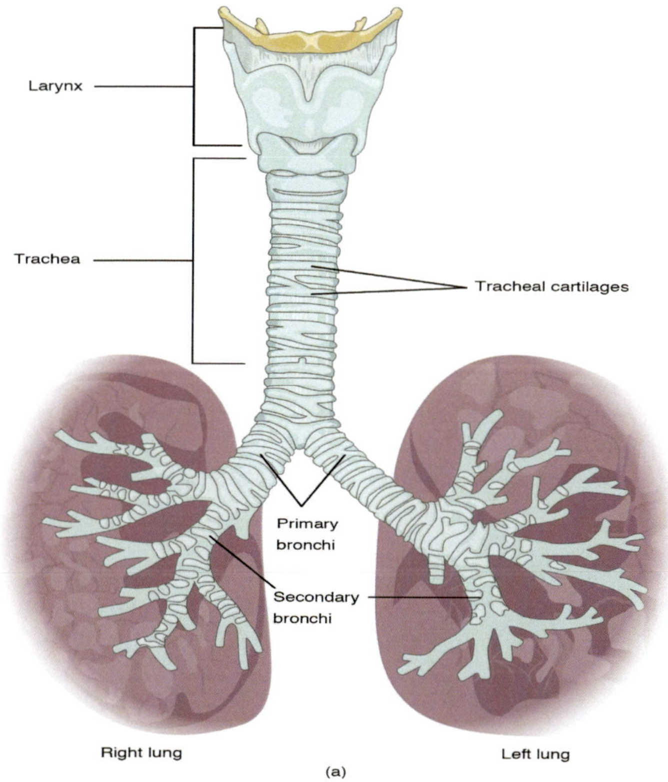

Figure 76. The Trachea.

Source : Image By Wikimedia Commons

Tracheal System

A system of branched, air-filled tubes that spreads throughout the body and transports oxygen directly to cells in insects.

Tracheid

A long, tapering water conducting cell found in nearly all vascular plants' xylem. Tracheids that work are no longer alive.

Trans Fat

An unsaturated fat with one or more trans double bonds that is generated artificially through the hydrogenation of oils.

Transduction

(1) The transfer of bacterial DNA by phages (viruses) from one bacterial cell to another. Transduction results in horizontal gene transfer when these two cells are members of different species. (2) In cellular communication, the transformation of a signal from outside the cell into a form capable of eliciting a specific cellular response; also known as signal transduction.

Transfer RNA (tRNA)

An RNA molecule that acts as a translator between nucleic acid and protein languages by transporting certain amino acids to the ribosome, where they identify the right mRNA codons.

Transformation

(1) The transformation of a healthy animal cell into a malignant cell. (2) A change in genotype and phenotype caused by a cell's assimilation of foreign DNA. Transformation results in horizontal gene transfer when the foreign DNA is from a different species.

Transgenic

Pertaining to an organism whose genome contains a gene introduced from another organism of the same or a different species.

Translation

An organism whose genome contains a gene introduced from another organism of the same or a different species.

Translocation

(1) A chromosomal structural abnormality caused by the attachment of a chromosomal fragment to a nonhomologous chromosome. (2) During protein synthesis, the third step of the elongation cycle, when the RNA transporting the developing polypeptide on the ribosome passes from the A site to the P site. (3) The movement of organic nutrients in vascular plant phloem.

Transmission Electron Microscope (TEM)

A microscope that studies the interior ultrastructure of cells by passing an electron beam through very thin sections stained with metal atoms.

Transport Protein

A transmembrane protein that aids in the passage of a certain drug or class of closely related molecules through the membrane.

Transposable Element

A transposable genetic element is a fragment of DNA that can travel within a cell's genome via a DNA or RNA intermediary.

Triiodothyronine (T3)

One of two iodine-containing hormones generated by the thyroid gland that aids in the regulation of metabolism, development, and maturation in vertebrates.

Triple Response

In reaction to mechanical stress, a plant growth manoeuvre that involves delaying stem elongation, thickening of the stem, and a curvature that leads the stem to begin growing horizontally.

Triplet Code

A genetic information system in which the amino acids for polypeptide chains are specified by a series of three-nucleotide- words.

Triploblastic

Endoderm, mesoderm, and ectoderm are the three germ layers. The majority of eumetazoans are triploblastic.

Trochophore Larva

Some lophotrochozoan animals, including annelids and mollusks, have a distinct larval stage.

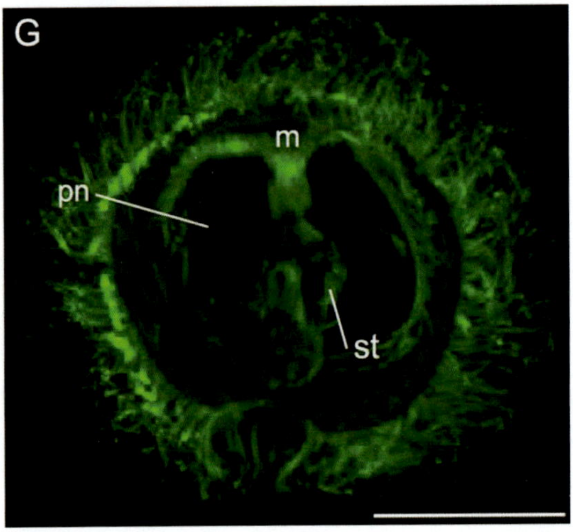

Figure 77. Complete Trochophore Larva of The Annelid Pomatoceros Lamarckii *Visualized* With Immnunohisto - chemistry With Anti-Acetylated Tubulin, Posterior View.

Source : Image By Wikimedia Commons

Trophic Structure

The various feeding interactions in an ecosystem that determine the energy flow path and chemical cycle pattern.

Trophoblast

A mammalian blastocyst's outer epithelium. It is the foetal component of the placenta, aiding embryonic development but not actually being a part of the embryo.

Tropical Dry Forest

A terrestrial biome with relatively high temperatures and precipitation overall, but a distinct dry season.

Tropism

A growth response characterised by the curvature of entire plant organs toward or away from stimuli as a result of varying rates of cell elongation.

True-Breeding

Describes animals that produces progeny of the same genotype over multiple generations of self-pollination.

Tubal Ligation

A method of sterilisation in which the two oviducts (fallopian tubes) of a woman are tied shut to prevent eggs from reaching the uterus. Each oviduct has a section removed.

Tube Foot

One of many water circulatory system extensions of an echinoderm. Tube feet help with locomotion and feeding.

Tumor-Suppressor Gene

A gene whose protein product suppresses cell division, preventing cancer-causing uncontrolled cell growth.

Tundra

A terrestrial biome at the apex of plant development. It is known as arctic tundra in the far north, and alpine tundra in the high mountains, where plant types are limited to low shrubby or matlike flora.

Figure 78. Snow Fields Lie on The Tundra at Rocky Mountain National Park.

Source : Image By Wikimedia Commons

Turgid

Plant cells are swollen or bloated. (A walled cell becomes turgid if its water potential is lower than that of its surroundings, allowing water to enter.)

Turgor Pressure

The force exerted against a plant cell wall as a result of water input and cell enlargement caused by osmosis.

Turnover Time

The time necessary to replace a populations or group of populations' standing crop (for example, phytoplankton), calculated as the standing crop to production ratio.

Twin Study

A behavioural study in which researchers compare the behaviour of identical twins raised in different households to that of identical twins grown in the same household.

U

UniGene Database

A public database, maintained by NCBI, which brings together sets of GenBank sequences that represent the transcription products of distinct genes.

Unikonta

One of five eukaryotic supergroups postulated in a contemporary hypothesis of eukaryotic evolution. This clade, which includes amoebozoans and opisthokonts, is supported by analyses of myosin proteins and DNA. Excavata, Chromalveolata, Rhizaria, and Archaeplastida are other genera.

Unsaturated Fatty Acid

A fatty acid with one or more double bonds in the hydrocarbon tail. The number of hydrogen atoms linked to the carbon skeleton is reduced as a result of such bonding.

Urea

A nitrogenous waste that is soluble and formed in the liver via a metabolic cycle that combines ammonia and carbon dioxide.

Urethra

A tube that excretes urine from the mammalian body near the vagina in females and through the penis in men; also functions as the reproductive system's exit route in males.

Uric Acid

A nitrogenous waste product of insects, land snails, and many reptiles that is produced as a byproduct of protein and purine metabolism. Uric acid is non-toxic and largely insoluble.

Uterine Cycle

The changes that occur in the uterus during the human female's reproductive cycle; also known as the menstrual cycle.

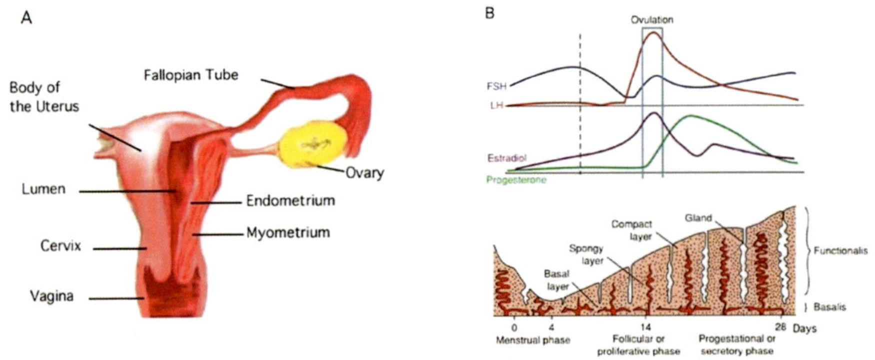

Figure 79. Uterine Anatomy.

Source : Image By Wikimedia Commons

Utricle

A vestibule chamber behind the oval window that leads into the three semicircular canals of the vertebrate ear.

V

Vaccine

A pathogen variation or derivative that encourages the immune system of the host to mount defences against the pathogen.

Vagina

The birth canal in mammals is a part of the female reproductive system that connects the uterus to the outside entrance. The vagina accommodates the male penis and receives sperm during copulation.

Valence

A given atom's bonding capacity is usually equal to the number of unpaired electrons necessary to complete the atom's outermost (valence) shell.

Valence Shell

The outermost energy shell of an atom, which contains the valence electrons involved in the atom's chemical processes.

Variable Numbers of Tandem Repeats (VNTRs)

DNA sequence blocks of 2 to 60 base pairs which are repeated from two to more than 20 times in different individuals. This polymorphism makes VNTRs very useful DNA markers used in genomic mapping, linkage analysis, and DNA fingerprinting.

Vas Deferens

The tube in the male reproductive system that transports sperm from the epididymis to the urethra in mammals.

Vascular Cambium

In woody plants, a cylinder of meristematic tissue that adds layers of secondary vascular tissue called secondary xylem (wood) and secondary phloem.

Vascular Plant

A plant that contains vascular tissue. Except for liverworts, mosses, and hornworts, vascular plants contain all living plant species.

Vascular Tissue

Plant tissue is made up of cells that are linked together to form tubes that transport water and nutrients throughout the plant's body.

Vascular Tissue System

A vascular plant's transport system composed of xylem and phloem. Water and minerals are transported by xylem, whereas sugars, the results of photosynthesis, are transported by phloem.

Vaso-Congestion

The filling of a tissue with blood as a result of increased blood flow through that tissue's arteries.

Vegetal Pole

The region at the end of an egg in a hemisphere where the majority of the yolk is concentrated; the opposite of the animal pole.

Version

Similar to the Protein ID for protein sequences, the version is a nucleotide sequence identification number assigned to each GenBank sequence. The format for this sequence identifier is accession. version (e.g., M12345.1). Whenever the author of a particular sequence record changes the sequence data in any way (even if just a single nucleotide is altered), the version number will be increased by an increment of one while the accession number base remains constant. For example, M12345.1 would become M12345.2. Each sequence change also results in the assignment of a new GI number (link to GI entry). Whenever an NCBI sequence database is searched, only the most recent version of a record is retrieved. NCBI's Sequence Revision History page is used to view the various GI numbers, version numbers, or update dates associated with a particular GenBank record.

Ventricle

(1) A cardiac chamber in which blood is pumped out of the heart. (2) A cerebrospinal fluid-filled compartment in the vertebrate brain.

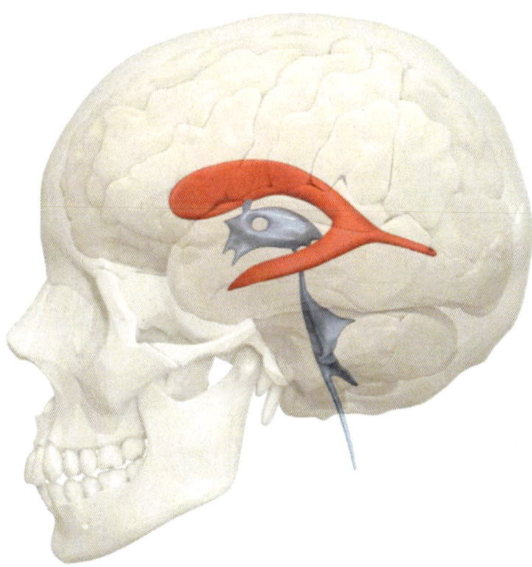

Figure 80. Lateral Ventricle.

Source : Image By Wikimedia Commons

Vertebrate

Sharks and rays, ray-finned fishes, coelacanths, lungfishes, amphibians, reptiles, and mammals are all chordate animals having backbones.

Vessel

Most angiosperms and a few nonflowering vascular plants have a continuous water-conducting micropipe.

Vessel Element

A water-conducting cell that is short and broad and present in the xylem of most angiosperms and a few nonflowering vascular plants. At maturity, vessel parts are placed end to end to form vessels, which are micropipes.

Vestigial Structure

A characteristic of an organism that is a historical relic of a structure that once served a purpose in the organism's forebears.

Villus

(1) A finger-like protrusion of the small intestine's inner surface. (2) A finger-like protrusion of the mammalian placenta's chorion. The presence of a large number of villi increases the surface area of these organs.

Viral Envelope

A membrane produced from host cell membranes that encases the capsid, which in turn encloses a viral genome.

Virtual Libraries

The creation and storage of vast collections of molecular structures in an electronic database. These databases may be queried for subsets that exhibit specific physicochemical features, or may be "virtually screened " for their ability to bind a drug target. This process may be performed prior to the synthesis and testing of the molecules themselves.

Virus

An infectious particle that cannot replicate outside of a cell and is composed of an RNA or DNA genome surrounded by a protein coat (capsid) and, in the case of some viruses, a membranous envelope.

Visible Light

The fraction of the electromagnetic spectrum that may be recognised by the human eye as different colours, with wavelengths ranging from about 380 nm to about 750 nm.

Visualization

A process of representing abstract scientific data as images that can aid in understanding the meaning of the data.

Vitamin

An chemical molecule that is required in very small levels in the diet. Many vitamins function as coenzymes or components of coenzymes.

Viviparous

Refers to a type of development in which the young are born alive after being nourished in the uterus by placental blood.

W

Water Potential (Ψ)

The physical property that predicts the direction of water flow, which is regulated by solute concentration and applied pressure.

Water Vascular System

A network of hydraulic canals that is unique to echinoderms and branches into tube feet that function in locomotion and eating.

Wetland

A habitat that is inundated by water at least some of the time and that supports plants adapted to water-saturated soil.

Figure 81. Freshwater Wetland, Hunter Wetlands National Park, Newcastle.

Source : Image By Flickr

Wild Type

The most prevalent phenotype observed in natural populations; also refers to the individual who has that trait.

Wobble

The ability of the nucleotide at the 5 end of a tRNA anticodon to establish hydrogen bonds with more than one type of base in the third position (3 end) of a codon.

X

X-Ray Crystallography

A method for studying the three-dimensional structure of molecules. It is determined by the X-ray beam diffraction by the individual atoms of a crystalline molecule.

Xylem

Vascular plant tissue is made up primarily of tubular dead cells that transport the majority of the water and minerals from the roots to the rest of the plant.

Y

Yeast

Fungus with one cell. Yeasts reproduce asexually through binary fission or by pinching off tiny buds from a parent cell. Many fungal species can grow both as yeasts and as filamentous networks; just a few species grow entirely as yeasts.

Z

Zero Population Growth (ZPG)

A period of population stability in which additions to the population from births and immigration are balanced by subtractions from deaths and emigration.

Zone Of Polarizing Activity (ZPA)

A mesodermal block positioned just under the ectoderm where the posterior side of a limb bud is linked to the body; needed for appropriate pattern formation along the limb's anterior-posterior axis.

Zoned Reserve

A large territory that comprises places that have been generally unaltered by humans but are bordered by areas that have been altered by human activity and are used for economic advantage.

Zygomycete

Zygomycota fungal phylum member distinguished by the creation of a strong structure termed a zygosporangium during sexual reproduction.

Zygosporangium

A strong multinucleate structure in zygomycete fungus where karyogamy and meiosis occur.

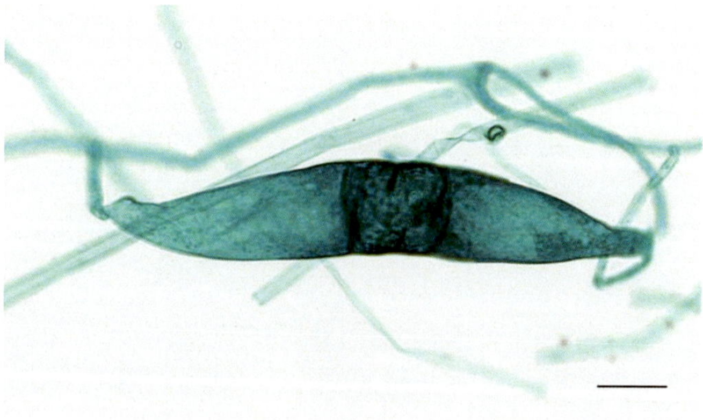

Figure 82. Light Microscopy of Rhizopus *Showing the Early Stages of* Plasmogamy When the Immature Zygosporangium Begins to Form.

Source : Image By Wikimedia Commons

Zygote

In zygomycete fungi, a strong multinucleate structure where karyogamy and meiosis occur.

REFERENCES

1. Higgins, D., 2005. Book Review: Dictionary of Bioinformatics and Computational Biology. By Hancock, John M. and Zvelebil, Marketa J. (Eds). *PROTEOMICS*, [online] 5(7), pp.1996-1996. Available at: <https://analyticalsciencejournals.onlinelibrary.wiley.com/doi/10.1002/pmic.200590022> [Accessed 10 May 2022].
2. Madhavan, G., 2005. Dictionary of Bioinformatics and Computational Biology. Edited by John M. Hancock and Marketa J. Zvelebil, ISBN: 0-471-43622-4, Hardcover, 664 pages, July 2004, US $100. *Annals of Biomedical Engineering*, [online] 33(3), pp.413-413. Available at: <https://link.springer.com/article/10.1007/s10439-005-1744-8> [Accessed 10 May 2022].
3. Madhavan, G., 2005. Dictionary of Bioinformatics and Computational Biology. Edited by John M. Hancock and Marketa J. Zvelebil, ISBN: 0-471-43622-4, Hardcover, 664 pages, July 2004, US $100. *Annals of Biomedical Engineering*, [online] 33(3), pp.413-413. Available at: <https://link.springer.com/article/10.1007/s10439-005-1744-8> [Accessed 10 May 2022].
4. Marchet, C., Lecompte, L., Limasset, A., Bittner, L. and Peterlongo, P., 2020. A resource-frugal probabilistic dictionary and applications in bioinformatics. *Discrete Applied Mathematics*, [online] 274, pp.92-102. Available at: <https://www.sciencedirect.com/science/article/abs/pii/S0166218X18301288?via%3Dihub> [Accessed 10 May 2022].
5. Oyawoye, O., Olorunsola, H., Oluduro, A., Bamigboye, O., Oladipo, E., Olotu, T., Adeosun, I., Kaka, M. and Obembe, O., 2022. Bioinformatics. *Medical Biotechnology, Biopharmaceutics, Forensic Science and Bioinformatics*, [online] pp.213-223. Available at: <https://

www.taylorfrancis.com/chapters/edit/10.1201/9781003178903-13/bioinformatics-oyawoye-olorunsola-oluduro-bamigboye-oladipo-olotu-adeosun-kaka-obembe> [Accessed 10 May 2022].

6. Pegg, S., 2005. Dictionary of Bioinformatics and Computational Biology. *Briefings in Bioinformatics*, [online] 6(2), pp.211-212. Available at: <https://academic.oup.com/bib/article/6/2/211/249578?login=false> [Accessed 10 May 2022].
7. Saussy, H., 2021. Bioinformatics. *Information*, [online] pp.57-71. Available at: <https://www.degruyter.com/document/doi/10.7312/kenn19876-006/html> [Accessed 10 May 2022].
8. Schneider, G., 2005. Dictionary of Bioinformatics and Computational Biology. Edited by John M. Hancock and Marketa J. Zvelebil. *ChemBioChem*, [online] 6(7), pp.1287-1287. Available at: <https://chemistry-europe.onlinelibrary.wiley.com/doi/10.1002/cbic.200500190> [Accessed 10 May 2022].
9. Shuangge Ma, 2007. Book review: Hancock JM and Zvelebil MJ 2004. Dictionary of Bioinformatics and Computational Biology. Wiley-Liss. *Statistical Methods in Medical Research*, [online] 17(2), pp.223-223. Available at: <https://journals.sagepub.com/doi/10.1177/0962280207085807> [Accessed 10 May 2022].
10. Author, P., 2006. Management of Genetic Syndromes • Mapping our Ancestors: Phylogenetic Approaches in Anthropology and Prehistory • Dictionary of Bioinformatics and Computational Biology • Biology at Work: Rethinking Sexual Equality • The Adventure of English: The Biography of a Language. *Mankind Quarterly*, [online] 46(3), pp.0-0. Available at: <https://mankindquarterly.org/archive/issue/46-3/4> [Accessed 10 May 2022].
11. Izadkhah, H., 2022. Classification in bioinformatics. *Deep Learning in Bioinformatics*, [online] pp.113-130. Available at: <https://www.sciencedirect.com/science/article/pii/B9780128238226000135?via%3Dihub> [Accessed 10 May 2022].
12. Marques Sousa, J., Ribeiro Viegas, M., Teixeira Ferreira, P., Coelho Paiva, P., Oliveira, A., Fernandes, P. and Ramos, M., 2021. MOLECULAR BIOINFORMATICS. *Revista de Ciência Elementar*, [online] 9(4). Available at: <https://rce.casadasciencias.org/rceapp/static/docs/extras/molecularbioinformatics.pdf> [Accessed 10 May 2022].

13. Ötleş, S., Bakar, B. and Türköz, B., 2022. Bioinformatic Analysis. *Bioactive Peptides from Food*, [online] pp.321-346. Available at: <http://10.1201/9781003106524-20> [Accessed 10 May 2022].
14. Pal, A., 2021. Introduction to Bioinformatics. *Springer Protocols Handbooks*, [online] pp.129-158. Available at: <https://link.springer.com/protocol/10.1007/978-1-0716-1818-9_7> [Accessed 10 May 2022].
15. PRISACARU, A. and BUTNARU, D., 2022. ON THE CHALLENGES OF A DICTIONARY OF TOPONYMY. *Studii și cercetări de onomastică și lexicologie*, [online] 28(1-2), pp.145-157. Available at: <https://revistascol.ro/images/site_ro/2021/007_p145.pdf> [Accessed 10 May 2022].
16. Sofi, M., Shafi, A. and Masoodi, K., 2022. BioEdit in bioinformatics. *Bioinformatics for Everyone*, [online] pp.231-236. Available at: <https://linkinghub.elsevier.com/retrieve/pii/B9780323911283000227> [Accessed 10 May 2022].
17. Sofi, M., Shafi, A. and Masoodi, K., 2022. Prologue to bioinformatics. *Bioinformatics for Everyone*, [online] pp.1-7. Available at: <https://www.sciencedirect.com/science/article/pii/B9780323911283000185?via%3Dihub> [Accessed 10 May 2022].
18. Tiwary, B., 2021. Agricultural Bioinformatics. *Bioinformatics and Computational Biology*, [online] pp.183-201. Available at: <https://link.springer.com/chapter/10.1007/978-981-16-4241-8_10> [Accessed 10 May 2022].
19. Tiwary, B., 2021. Clinical Bioinformatics. *Bioinformatics and Computational Biology*, [online] pp.163-182. Available at: <https://link.springer.com/chapter/10.1007/978-981-16-4241-8_9> [Accessed 10 May 2022].
20. Tiwary, B., 2021. Structural Bioinformatics. *Bioinformatics and Computational Biology*, [online] pp.65-86. Available at: <https://link.springer.com/chapter/10.1007/978-981-16-4241-8_5> [Accessed 10 May 2022].
21. Chen, M., Hofestädt, R. and Taubert, J., 2022. Integrative Bioinformatics: History and Perspective. *Integrative Bioinformatics*, [online] pp.3-9. Available at: <https://link.springer.com/chapter/10.1007/978-981-16-6795-4_1> [Accessed 10 May 2022].

22. Fajfar, T. and Žagar Karer, M., 2022. Doublets in terminological dictionaries. *Pravopis na zrnu graha*, [online] Available at: <https://omp.zrc-sazu.si/zalozba/catalog/book/1992> [Accessed 10 May 2022].
23. Grealey, J., Lannelongue, L., Saw, W., Marten, J., Méric, G., Ruiz-Carmona, S. and Inouye, M., 2022. The Carbon Footprint of Bioinformatics. *Molecular Biology and Evolution*, [online] 39(3). Available at: <https://academic.oup.com/mbe/article/39/3/msac034/6526403?login=false> [Accessed 10 May 2022].
24. He, C. and Duan, Y., 2021. Bioinformatics. *Clinical Molecular Diagnostics*, [online] pp.45-59. Available at: <https://link.springer.com/chapter/10.1007/978-981-16-1037-0_5> [Accessed 10 May 2022].
25. Khadiev, K. and Remidovskii, V., 2021. Classical and Quantum Algorithms for Assembling a Text from a Dictionary. *Nonlinear Phenomena in Complex Systems*, [online] 24(3), pp.207-221. Available at: <http://www.j-npcs.org/abstracts/vol2021/v24no3/v24no3p207.html> [Accessed 10 May 2022].
26. Khadiev, K. and Remidovskii, V., 2021. Classical and quantum algorithms for constructing text from dictionary problem. *Natural Computing*, [online] 20(4), pp.713-724. Available at: <https://link.springer.com/article/10.1007/s11047-021-09863-1> [Accessed 10 May 2022].
27. not provided, b., 2022. PROTAC Design based on Bioinformatics v1. [online] Available at: <https://www.protocols.io/view/protac-design-based-on-bioinformatics-dm6gpbox1lzp/v1> [Accessed 10 May 2022].
28. not provided, b., 2022. PROTAC Design based on Bioinformatics v1. [online] Available at: <https://www.protocols.io/view/protac-design-based-on-bioinformatics-dm6gpbox1lzp/v1> [Accessed 10 May 2022].
29. Singh, R. and DUTTA, S., 2021. What is Bioinformatics. [online] Available at: <https://osf.io/vb52a/> [Accessed 10 May 2022].
30. Teikari, J., Baunach, M. and Dittmann, E., 2022. Cyanobacterial Genome Sequencing, Annotation, and Bioinformatics. *Methods in Molecular Biology*, [online] pp.269-287. Available at: <https://link.springer.com/protocol/10.1007/978-1-0716-2273-5_14> [Accessed 10 May 2022].

31. 2021. Computation in Bioinformatics. [online] Available at: <https://onlinelibrary.wiley.com/doi/book/10.1002/9781119654803> [Accessed 10 May 2022].
32. 2022. Bioinformatics for Everyone. [online] Available at: <https://www.sciencedirect.com/book/9780323911283/bioinformatics-for-everyone?via=ihub=> [Accessed 10 May 2022].
33. Abed, R. and Al-Najjar, Y., 2021. Bioinformatics Storing Databases. *Technium BioChemMed*, [online] 2(4), pp.96-105. Available at: <https://techniumscience.com/index.php/biochemmed/article/view/5335> [Accessed 10 May 2022].
34. Bello-Chavolla, O., Fernández-Chirino, L., Antonio-Villa, N., Delaye-Martinez, M. and Sicilia-Andrade, A., 2022. Bioinformatics and Genomics for Epidemiologists. *Principles of Genetics and Molecular Epidemiology*, [online] pp.131-143. Available at: <https://link.springer.com/chapter/10.1007/978-3-030-89601-0_12> [Accessed 10 May 2022].
35. Gupta, M. and Behera, L., 2021. Introduction to Bioinformatics. *Bioinformatics in Rice Research*, [online] pp.3-20. Available at: <http://10.1007/978-981-16-3993-7_1> [Accessed 10 May 2022].
36. Gupta, M., Gouda, G., Donde, R., Sabarinathan, S., Dash, G., Ponnana, M., Pati, P., Rathore, S., Vadde, R. and Behera, L., 2021. Statistics for Bioinformatics. *Bioinformatics in Rice Research*, [online] pp.21-50. Available at: <https://link.springer.com/chapter/10.1007/978-981-16-3993-7_2> [Accessed 10 May 2022].
37. Mooney, S., Tenenbaum, J. and Altman, R., 2021. Bioinformatics. *Biomedical Informatics*, [online] pp.273-298. Available at: <https://link.springer.com/chapter/10.1007/978-3-030-58721-5_9> [Accessed 10 May 2022].
38. Shah, S., Hameed, A., Ali Almazroi, A. and Alqarni, M., 2022. Agent-Based Data Extraction in Bioinformatics. *Security and Interaction Networks*, [online] 2022, pp.1-11. Available at: <https://www.hindawi.com/journals/scn/2022/4865209/> [Accessed 10 May 2022].
39. Sofi, M., Shafi, A. and Masoodi, K., 2022. Bioinformatics databases and tools. *Bioinformatics for Everyone*, [online] pp.17-23. Available at: <https://www.sciencedirect.com/science/article/pii/B9780323911283000094?via%3Dihub> [Accessed 10 May 2022].

40. White, S., Heijne, G. and Engelman, D., 2021. Membrane Protein Bioinformatics. *Cell Boundaries*, [online] pp.265-290. Available at: <https://www.taylorfrancis.com/chapters/mono/10.1201/9780429341328-9/membrane-protein-bioinformatics-stephen-white-gunnar-von-heijne-donald-engelman> [Accessed 10 May 2022].

INDEX

I

K

L

M

N

O

P

R

S

T

U

V

W

X

Y

FSC
www.fsc.org
MIX
Papier | Fördert gute Waldnutzung
FSC® C083411

Kolibri 360 GmbH
Ferdinand-Jühlke-Straße 7
99095 Erfurt, Deutschland
produktsicherheit@kolibri360.de